Paraffin - Thermal Energy Storage Applications

Edited by ElSayed G. Zaki and Abdelghaffar S. Dhmees

Published in London, United Kingdom

IntechOpen

Supporting open minds since 2005

Paraffin - Thermal Energy Storage Applications
http://dx.doi.org/10.5772/intechopen.90983
Edited by ElSayed G. Zaki and Abdelghaffar S. Dhmees

Contributors
Hebatallah Teamah, Eloi Alves da Silva Filho, Fabrício Uliana, Arlan da Silva Gonçalves, Hani Hussain Hussain Sait, Maher Mohammad Al-Maghalseh, Gulfam Raza, Saqib Iqbal, Abdul Samad Farooq, ElSayed G. Zaki, Shimaa Elsaeed, Abdelghaffar S. Dhmees

First published in London, United Kingdom, 2022 by IntechOpen
IntechOpen is the global imprint of INTECHOPEN LIMITED, registered in England and Wales, registration number: 11086078, 5 Princes Gate Court, London, SW7 2QJ, United Kingdom
Printed in Croatia

British Library Cataloguing-in-Publication Data
A catalogue record for this book is available from the British Library

Additional hard and PDF copies can be obtained from orders@intechopen.com

Paraffin - Thermal Energy Storage Applications
Edited by ElSayed G. Zaki and Abdelghaffar S. Dhmees
p. cm.
Print ISBN 978-1-83968-705-1
Online ISBN 978-1-83968-706-8
eBook (PDF) ISBN 978-1-83968-707-5

For EU product safety concerns:
IN TECH d.o.o., Prolaz Marije Krucifikse Kozulić 3, 51000 Rijeka, Croatia, info@intechopen.com or visit our website at intechopen.com.

Meet the editors

Dr. ElSayed G. Zaki received his Ph.D. in Physical Chemistry from the Department of Chemistry, Faculty of Science, Mansoura University, Egypt. Currently, he is a visiting researcher at the Department of Chemical and Biomolecular Engineering, Iacocca Hall, Lehigh University, Pennsylvania, USA. He is an Associate Professor of Applied Chemistry in the Petroleum Applications Department, Egyptian Petroleum Research Institute. He is currently an editorial board member for several reputed journals, a member of the Synchrotron-light for Experimental Science and Applications in the Middle East (SESAME), Organization for Women in Science for the Developing World (OWSD), and the International Union of Pure and Applied Chemistry (IUPAC).

Dr. Abdelghaffar S. Dhmees has been working as an assistant professor at the Department of Analysis & Evaluation, Egyptian Petroleum Research Institute, Egypt. He obtained a Ph.D. in 2015 with a thesis on the valorization of industrial waste to heterogeneous catalysts for biodiesel production. He is currently working on the synthesis of advanced 2D materials (e.g., MoS2, graphene, LDH, and its composites) for application in energy storage and valorization of industrial waste and byproducts for the synthesis of adsorbents, catalysts, membranes, and photocatalysts for application in water treatment, petroleum stream purification, and biofuel production.

Contents

Preface **XI**

Chapter 1 **1**
Advancement in Energy Storage by Paraffin
by ElSayed G. Zaki, Shimaa M. Elsaeed and Abdelghaffar S. Dhmees

Chapter 2 **7**
Latent Heat Storage: An Introduction
by Hebatallah Teamah

Chapter 3 **19**
Paraffin Wax-Based Thermal Composites
by Gulfam Raza, Saqib Iqbal and Abdul Samad Farooq

Chapter 4 **37**
CFD Model of Shell-and-Tube Latent Heat Thermal Storage Unit Using Paraffin as a PCM
by Maher Mohammad Al-Maghalseh

Chapter 5 **61**
Solidification and Melting of Phase Change Material in Cold Thermal Storage Systems
by Hani Hussain Sait

Chapter 6 **85**
News Aspects Theoric and Experimental to Paraffins Compounds
by Eloi Alves da Silva Filho, Fabrício Uliana and Arlan da Silva Gonçalves

Preface

Paraffin oil and mineral oil are synonymous terms. Fluid paraffin oil is a mineral oil that is a byproduct of unrefined petroleum refining. It is a straightforward, lackluster, unscented, and dull oil, chiefly made from high-bubbling alkane subsidiaries. However, it has the potential as a low-cost energy storage material.

Different forms of energy can be stored, including mechanical, electrical, and thermal energy. Energy storage not only reduces the mismatch between supply and demand but also improves the performance and reliability of energy systems and plays an important role in conserving energy. Thermal energy can be stored during the off-peak period, usually at night, and re-used during the peak period.

This book focuses on the application of paraffin for thermal energy storage. It examines the preparation of paraffin via encapsulation to develop a nonconventional energy storage material.

ElSayed G. Zaki
Associate Professor,
Petroleum Applications Department,
Egyptian Petroleum Research Institute,
Cairo, Egypt

Abdelghaffar S. Dhmees
Associate Professor,
Analysis and Evaluation Department,
Egyptian Petroleum Research Institute,
Cairo, Egypt

Chapter 1

Advancement in Energy Storage by Paraffin

ElSayed G. Zaki, Shimaa M. Elsaeed
and Abdelghaffar S. Dhmees

Abstract

Paraffin uses in energy storage depends on preparation by encapsulation method become more effective nonconventional technique novel storge material. Many measurements as hydrophilicity, energy storage capacity, size distribution and encapsulation ratio can be evaluated. It was also found that a higher coating to paraffin ratio leads to a higher paraffin encapsulation ratio. The hydrophilicity value of microencapsulated paraffin depended mainly on the ratio of paraffin to coating the higher the ratio, the lower was its product hydrophilicity Surface response method used to design and based conditions to optimize it. Using paraffin in energy storage in the future is promising.

Keywords: energy storage, paraffin, surface response method, phase change material

1. Introduction

The main concern for the world is environmental sustainability challenge. Depending on fossil fuels and environmental pollution are the most driving force for the technologies of the future. Using renewable energy and its optimization is the vital factor of energy.

Energy storage is a vital part of energy saving and power supply, as energy demand and energy availability often does not coincide in time.

Storing thermal energy, sensible, latent and thermochemical energy storage are the three main ways. Paraffin oil or liquid paraffin oil is obtained in the process of crude oil distillation It is a colorless and odorless oil that is used for varied purposes. In some cases, *paraffin oil* and *mineral oil* are synonymous terms [1]. Fluid paraffin oil is a mineral oil and is a side-effect of unrefined petroleum refining. It is straightforward, lackluster, unscented, and dull oil, which is chiefly made out of high-bubbling alkane subsidiaries. Fluid paraffin (high-bubbling mineral oil) is a combination of higher sub-atomic weight alkane subordinates and has various names, including nujol.

2. Nanomaterials paraffin in erergy storge

Synthesis of new phase change materials is very important role in development of energy. So we were synthesis a series of blends for paraffin of solid solid phase change materials, side chains based on crystalline epoxy resin and diamine of poly propylene oxide together through a one-pot curing process [2].

The examination aftereffects of a novel nanoparticle-paraffin-following earthenware composite stage change material (NCPCM) for dormant warmth nuclear power stockpiling applications. The NCPCMs are created by unconstrained soften invasion of paraffin wax and profoundly conductive nanoparticles (e.g., nano-graphene) in a permeable earthenware system [3] numbers are utilized as an illustration to show the example preparation measure in our labs. To make bigger NCPCM units, the cycle can be increased. The iron ore tailing permeable media is created by a froth gel-projecting strategy as show in **Figure 1**.

An energy stockpiling framework has been intended to study the warm attributes of paraffin wax with an inserted nano size copper oxide (CuO) molecule [4].

Figure 2 shows the variation of thermal conductivity with respect to temperature for paraffin wax and nanoparticle—in paraffin wax emulsion of various concentrations. The exhibition improvement of structure stable PCMs with the expansion of peeled graphene nano-platelets (xGnP) as a warmth move advertiser [5]. Characterization of test to determine cross of nanomaterial's based on silicon oxide and cesium oxide nanoparticles on thermo-actual attributes of the paraffin through stage change material (PCM) [6].

The composite blend was sonicated for an hour utilizing a sonicator to get the fine scattering of the paraffin nanoparticles. The surfactant compound in design was not used in this examination to avoid their impact on the warm conductivity of the half and half nano/paraffin. The pre-arranged cross breed nano/paraffin tests were then shaped into the round and hollow structure as displayed in **Figure 3** [6].

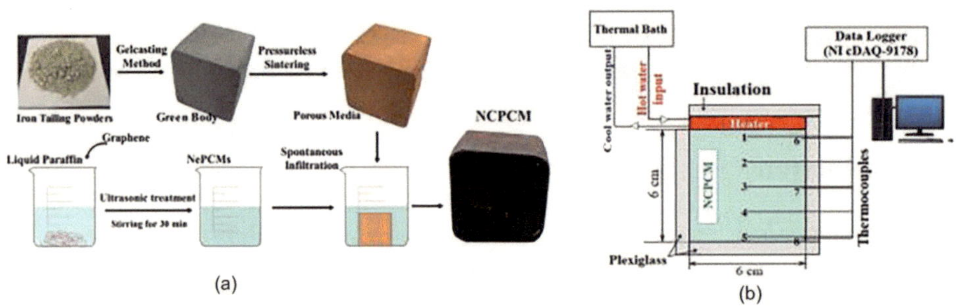

Figure 1.
(a) The NCPCM sample preparation process [3]. (b) Schematic experimental setup for enhanced heat transfer demonstration.

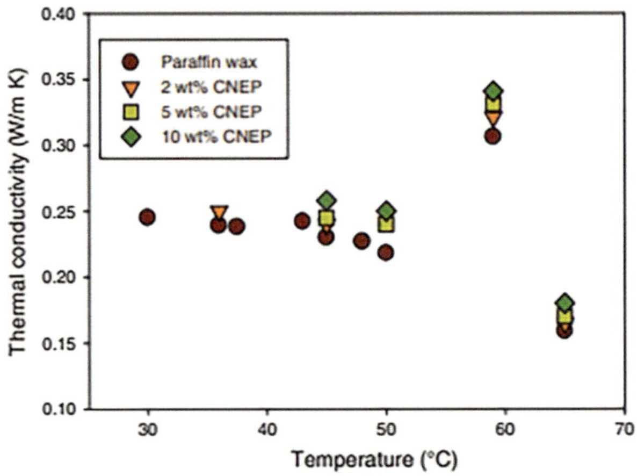

Figure 2.
Thermal conductivity of the paraffin wax and CuO—in paraffin wax [4].

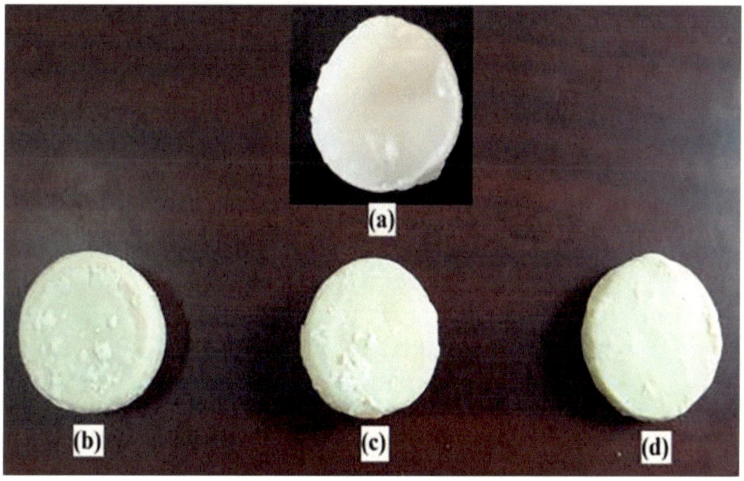

Figure 3.
Molded hybrid-nano/paraffin samples. (a) Pure paraffin; (b) 0.5 HnP; (c) 1.0 HnP; and (d) 2.0 [6].

Examining the warm properties of three concentrations vary from 0.5 to 1.5 wt% of titania-silver nanocomposite particles scattered paraffin wax PCM without and with sodium dodecyl sulfate (SDS) surfactant for both non-cycled and warm cycled tests. The colloidal arrangement is then splashed with different times to eliminate the wastes from the arrangement. The gotten arrangement is then separated, was dried at temperatures differ from 90 to 100°C for 12 hours and then dried to fine phase. Titania-silver NCP is delivered with the means are too represented in **Figure 4** [7].

Suspension polymerization method are used to form microcapsules containing paraffin wax as centers and polystyrene. Formtion of four exploratory elements, including level of initiator/styrene mass proportion (BPO/St wt.%), paraffin wax/styrene mass proportion (PCM/St), level of stabilizer/styrene mass proportion (PVP/St wt.%), and water/styrene mass proportion (H_2O/St), on microcapsules properties were researched [8].

Figure 5 displays the insertion of an and C on molecule extent. It tends to be seen that base molecule extent will accomplished at most extreme percent of BPO/St and PVP/St, despite the fact that at small degrees of %PVP/St the expanding of %BPO/St has a contrariwise impact on molecule extent.

Figure 6 shows the essential advances engaged with the emulsion polymerization strategy. In this strategy, an insoluble monomer contained in the dissolvable gets scattered consistently by the method for mechanical mixing in the response medium which contains a specific emulsifier and surfactant [9].

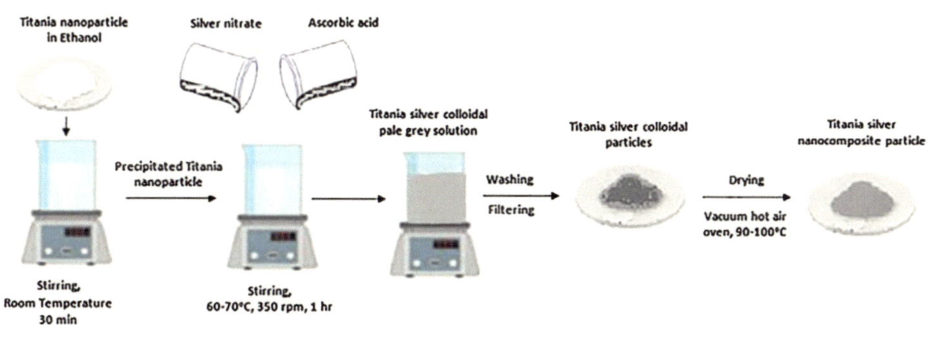

Figure 4.
Preparation of nanocomposite particles [7].

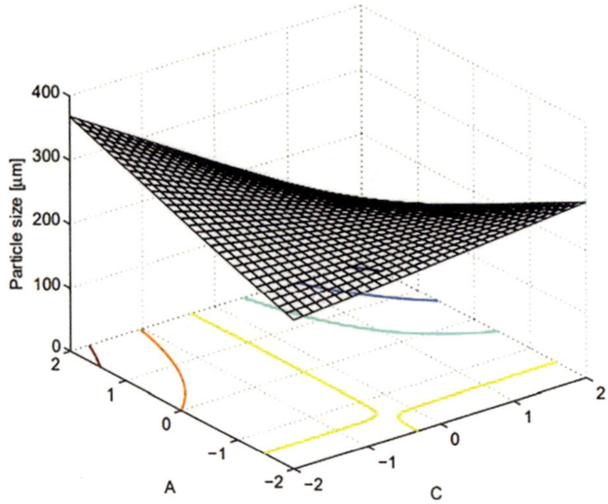

Figure 5.
Effect of a (%initiator/styrene mass proportion) and C (%stabilizer/styrene mass proportion) on particle size [8].

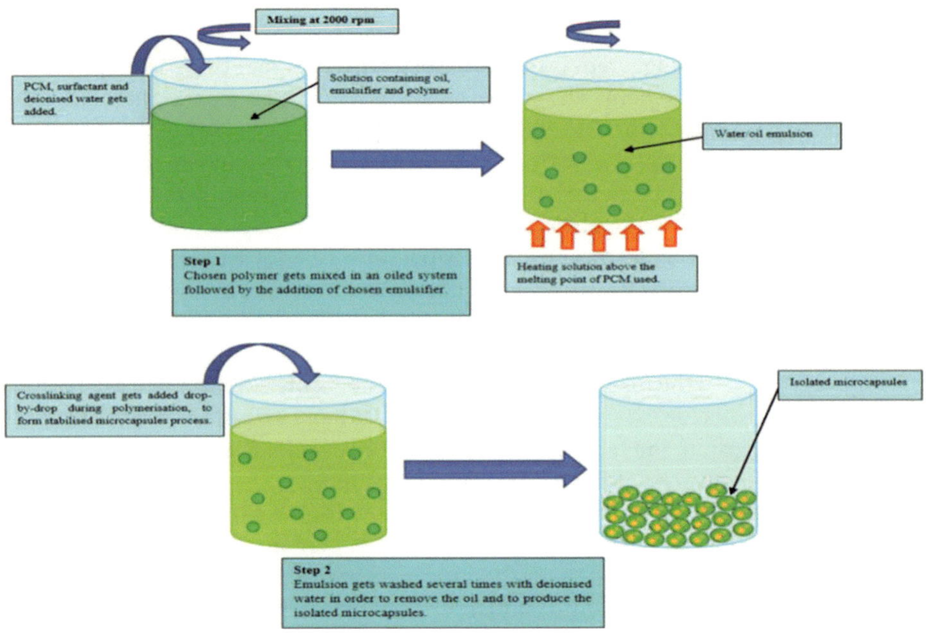

Figure 6.
Schematic representation of in situ polymerization method employed for the synthesis of microcapsules [9].

The EPDM rubber network features required for low leakage and good thermal performance were determined by Vulcanizing Paraffin wax (PW) PCM and Ethylene-Propylene-Diene-Monomer (EPDM) together using varied Benzoyl Peroxide concentrations. PCM systems were once popular. Vulcanized EPDM impregnation in high-temperature molten PW was used to create this product [10].

Quantitatively investigates a horizontal finned shell and tube LTES unit with various triplex-layer PCM parameters along the radial direction. The impacts of metal fin arrangements and PCM parameters on the melting performance of the LTES unit are analyzed using a comprehensive storage density evaluation (CSDE) criterion, and the optimum structure is determined using the CSDE criteria [11].

3. Conclusion

Paraffin uses in energy storage are now very important role of paraffin to overcome shortage of energy. Nanoparticles paraffin in energy storage become more advancement in energy storage. Many materials are used in energy storage as Phase Charge materials by mixing sodium dodecyl sulfate (SDS) surfactant, titania-silver nanocomposite particles scattered paraffin wax and nano size copper oxide. Response surface methodology RSM used to determine suitable position in the required space.

Acknowledgements

This work has been supported by the Egyptian Petroleum Research Institute (EPRI) research.

Conflict of interest

The authors declare no conflict of interest.

Author details

ElSayed G. Zaki*, Shimaa M. Elsaeed and Abdelghaffar S. Dhmees
Egyptian Petroleum Research Institute, El Zohour Region, Nasr City, Cairo, Egypt

*Address all correspondence to: chemparadise17@yahoo.com

References

[1] Speight JG. Handbook of Industrial Hydrocarbon Processes. MA, United States: Gulf Professional Publishing; 2019

[2] Lian Q, Li Y, Sayyed AAS, Cheng J, Zhang J. Facile strategy in designing epoxy/paraffin multiple phase change materials for thermal energy storage applications. ACS Sustainable Chemistry & Engineering. 2018;**6**(3):3375-3384

[3] Li R, Zhou Y, Duan X. Nanoparticle enhanced paraffin and tailing ceramic composite phase change material for thermal energy storage. Sustainable Energy and Fuels. 2020;**4**(9):4547-4557

[4] Jesumathy S, Udayakumar M, Suresh S. Experimental study of enhanced heat transfer by addition of CuO nanoparticle. Heat and Mass Transfer. 2012;**48**(6):965-978

[5] Ramakrishnan S, Wang X, Sanjayan J, Wilson J. Heat transfer performance enhancement of paraffin/expanded perlite phase change composites with graphene nano-platelets. Energy Procedia. 2017;**105**:4866-4871

[6] Pasupathi MK, Alagar K, Mm M, Aritra G. Characterization of hybrid-nano/paraffin organic phase change material for thermal energy storage applications in solar thermal systems. Energies. 2020;**13**(19):5079

[7] Bose P, Amirtham VA. Effect of titania-silver nanocomposite particle concentration and thermal cycling on characteristics of sodium dodecyl sulfate added paraffin wax thermal energy storage material. Energy Storage. 2021;**3**(2):e192

[8] Jamekhorshid A, Sadrameli SM, Bahramian AR. Process optimization and modeling of microencapsulated phase change material using response surface methodology. Applied Thermal Engineering. 2014;**70**(1):183-189

[9] Sivanathan A et al. Phase change materials for building construction: An overview of nano/micro-encapsulation. Nanotechnology Reviews. 2020;**9**(1): 896-921

[10] Dashtizadeh Z, Abdeali G, Bahramian AR, Alvar MZ. Enhancement of thermal energy absorption/storage performance of paraffin wax (PW) phase change material by means of chemically synthesized Ethylene Propylene Diene Monomer (EPDM) rubber network. Journal of Energy Storage. 2022;**45**:103646

[11] Xu H, Wang N, Zhang C, Qu Z, Cao M. Optimization on the melting performance of triplex-layer PCMs in a horizontal finned shell and tube thermal energy storage unit. Applied Thermal Engineering. 2020;**176**:115409

Chapter 2

Latent Heat Storage: An Introduction

Hebatallah Teamah

Abstract

This chapter includes an introduction to thermal energy storage systems. It lists the areas of application of the storage. It also includes the different storage systems; sensible, latent, and chemical. It concentrates on the concept and the application of latent thermal storage. A detailed overview of the energy storage capacity of latent systems is discussed. The motivation and the challenge to incorporate phase change materials in the storage system are highlighted. Next, a classification of different phase change materials (PCMs) and their applicability in different temperature ranges of operations are analyzed. A thorough review will be presented for its industrial applications.

Keywords: storage, latent, phase change material, organic, inorganic, solar

1. Introduction

The aim of the current chapter is to provide the reader with basics related to thermal energy storage. It highlights the need for storage, different types of storage, and the applicability of each. It mainly focuses on the latent heat storage from the prospective of its integration to different applications. It includes a comprehensive summary for different phase change material classifications. It articulates the efforts that has been reported within different scholarly articles in the research community.

2. The need for thermal storage and its types

Thermal energy storage is inevitably needed when the energy source is characterized by its intermittency. For example, it is crucial for a solar thermal system. **Figure 1** shows how the solar irradiation curve typically looks like. It shows the incident solar radiation, the useful collected solar gain, and the load if this system is used in residential heating applications. The useful collected gain is less than incident radiation because of collector absorbance properties. The figure shows that there are times of the day where there is excess solar supply relative to the demand. However, there are other times where there is deficient supply relative to the demand. The storage in this case is indispensable to even out this mismatch between the supply and the demand. It ensures that the residential demand is supplied whenever needed. The optimization of needed storage sizing is a tedious process. It needs to take into account a robust control mechanism to allow for stable operation.

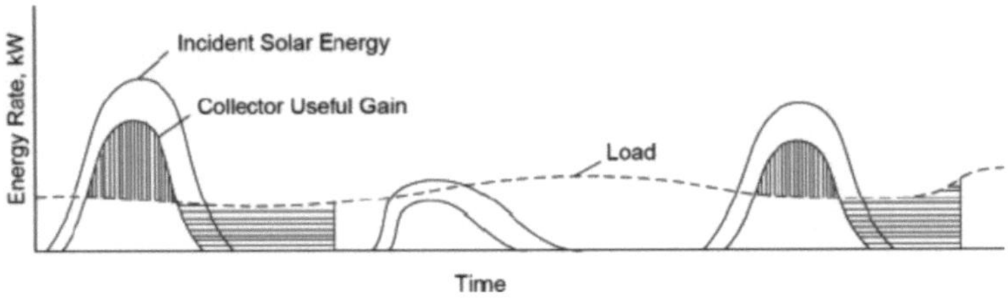

Figure 1.
Mismatch between supply and demand [1].

Energy storage systems have numerous classifications in literature. The most common one is to classify it to the broad category of thermal and chemical storage (**Figure 2**) [2]. The thermochemical storage stores heat as a part of chemical reaction. This kind of storage is out of scope of this book. Our focus is directed towards the thermal storage. It is subcategorized into the sensible, and the latent types. For the sensible storage, storage material preserves its condition as a solid or a liquid. The stored energy is manifested through the sensible increase in temperature of the material. The most common sensible storage material are water and rocks. On the other hand, latent storage is mainly dependent on phase change from solid to liquid and vice versa. Phase change materials (PCMs) change their phase at constant temperature (melting or solidification temperature). It stores the heat as the latent heat of change in phase is very high compared to the sensible heat. The temperature range of operation is important to choose the proper system. Sensible system shows an advantage with the wider temperature range. Latent system outperforms the sensible one in the narrow ranges of operation.

To show the difference in energy storage capacity between sensible and latent storage. Two storage media are chosen; water as a sensible medium, and lauric acid as a latent medium. Lauric acid changes its phase at 42°C. **Figure 3** shows a comparison of energy storage density between them when different operating temperature ranges are considered [1]. The first considered range is a narrow one of 10°C, and the second is a wide range of 40°C. In the wide range of operation, the energy stored in lauric acid is 70% higher than that of water. In the narrow operating range, the energy stored in lauric acid increases to 400% relative to water. This shows that phase change materials are more beneficial in narrow operating ranges.

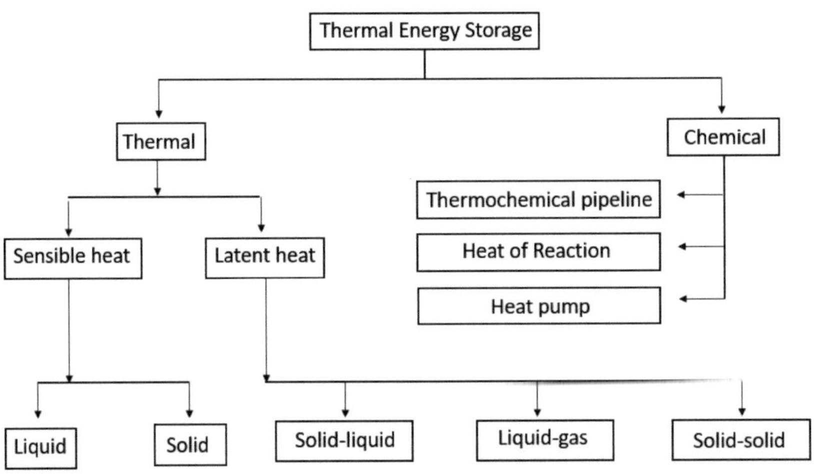

Figure 2.
Different thermal energy storage systems [2].

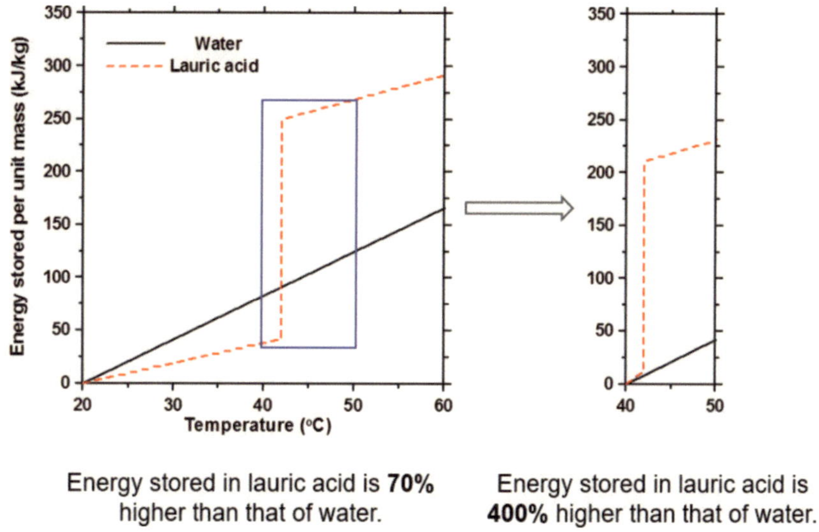

Figure 3.
Energy stored in lauric acid compared to water [1].

Especially as most of them possess poor thermal properties (density, and specific heat capacity). Latent heat storage attracted the research consideration through the past four decades. During the energy crisis at that time, PCMs were extensively studied in residential heating applications. There has been also a lot of efforts to maximize the benefits of latent storage and reduce the challenges that face its wide implication.

The special feature of high storage density of PCM, increased its integration in vast range of applications. Those applications include electronic cooling, smart buildings, and waste heat recovery. Most recently it was included in the water and space heating applications [3–9]. A typical heating system needs a stable temperature to maintain a good level of indoor comfort conditions. This can be achieved when latent storage is considered.

PCM classification is given in **Figure 4**. Different combination of probable phase change are shown. It might be from solid to liquid and vice versa as a common mode [10]. They are further subcategorized to; organic, inorganic, and composites. The inorganics have been dominant in the residential applications. They are less toxic and less corrosive. Their density does not change that much with phase change.

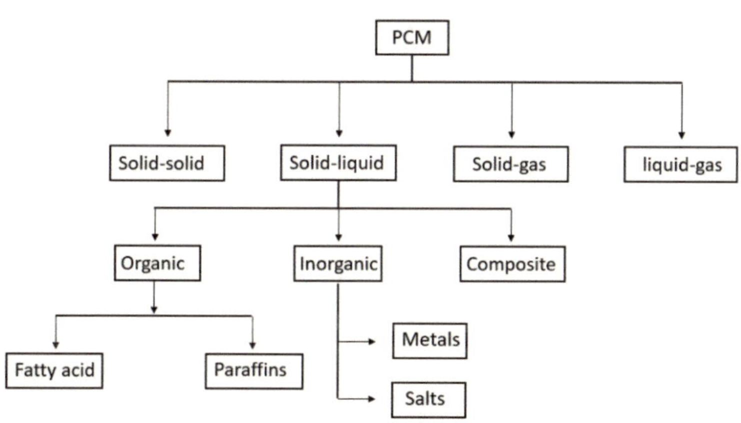

Figure 4.
PCM classification [2].

However, they are relatively more expensive with lower thermal conductivity. A family of organic fatty acids has been common as their melting point covers a wide temperature range (from 16–65°C). The inorganics especially salts are used in high temperature applications like solar concentrators.

Water is the most common sensible storage. A good way to make sense of PCM thermal properties it to compare them to water. Organic PCMs possess a specific heat capacity that is around 50% of that of water. The density is almost 80%. Their average latent heat of transformation is ~150 MJ / m^3. For inorganic PCM, the density is ~60% more than water. The specific heat capacity is half of water. However, the latent heat of fusion can be double the corresponding of an organic PCM.

Paraffins are considered a promising candidate for phase change materials. It is a hydrocarbon of high molecular mass. The melting point of paraffins differ with the number of atoms that can range from 12 to 40. The melting temperature includes a vast range from 6–80°C. This makes it suitable for several applications due to the wide range of melting temperature.

Paraffins have a high thermal storage capacity and the material freezes with marginal supercooling. It is highly stable under numerous cycles of melting and freezing. It is considered non corrosive in addition it is non-reactive to insulation material. This makes its encapsulation a straight forward process.

Hybrid thermal storage systems have shown a great promise in different applications. Those systems contain both water and PCMs. They combine the advantages of sensible and latent media. They also minimize the disadvantages of both of them. It should be taken into account that the configuration of heat exchanger is crucial for efficient energy storage. In addition, the PCM encapsulation should be carefully designed especially if the chosen PCM is of low conductivity. Spherical and cylindrical PCM encapsulations have shown a superior performance relative to rectangular ones [1].

The most common type of heat exchangers in hybrid storage is the shell and tube. Some researchers put the PCM in tubes and others put it in the shell. If water in the hybrid system is to be used in domestic applications, it is common to put PCM in tubes. Experiments can be done on this configuration in addition to numerical models. The numerical models account for phase change using different approaches.

The phase change process is tricky to model. The numerical model is nonlinear and transient throughout the process. Dutil et al. [11] summarized various numerical methods that has been reported to model phase change. The first most common one is the fixed grid. Such method presents a solution on unchanged grid through the entire numerical solution process. The second method that is not that common is the adaptive mesh. In such method, the mesh is varying throughout the solution. In the common fixed grid method, there are two subcategories. Those include the enthalpy porosity and specific heat capacity. The enthalpy porosity method is considered more stable. It adds an enthalpy term that is function of phase change percent. The main challenge of this method is to account for stagnation that that happens due to the solidification. This have been remedied by introducing the liquid fraction term [12]. This fraction is equal to unity when PCM is fully liquid and zero when PCM is solidified. It ranges from 0 to 1 throughout the rest of the process. The solution gets even trickier when convection dominated melting is considered [13–15]. There has been few approaches to take it into account. The straightforward one is the concept of equivalent thermal conductivity. Correlations are deployed to calculate its value. The higher the effect of convection, the higher the value of equivalent thermal conductivity. This causes it to deviate from the PCM conductivity and makes conduction dominated melting an invalid assumption.

The second fixed grid method is the heat capacity method. The heat of fusion is taken into account as a large effective heat capacity in a narrow phase change range [16–20]. This numerical method is easier to code relative to the enthalpy porosity method. However, there is numerical instability that arises from the narrow temperature range and selected integration procedure. This causes a difficulty in getting a converged solution. This has been overcome by using commercial software packages like COMSOL. They have developed a robust algorithm to account for the phase transition using the heat capacity method.

The configuration of latent and hybrid storage differs according to the application. The most common one is the shell and tube configuration. PCM can be put in either shell or tube. When the system is used in water heating application, PCM is put in tubes. Teamah et al. [21, 22] provided a detailed modeling for the operation of a shell and tube system (**Figure 5**). There is a parallel axial flow of water along PCM tubes. Melting considers both conduction and convection using correlations. One form to assess the feasibility of PCM integration is to compare energy storage gain. It is the stored energy in a hybrid system relative to a water based system. A nondimensional map has been concluded that is a function of design and operating parameters. The nondimensional map (**Figure 6**) shows that when Stephan number increases in wide temperature ranges, the system gains decrease. Gains also decrease if the heat capacity of PCM is lower than water. Better gain can be fulfilled if the charging period is longer. Similarly, it is higher if the thermal resistance of PCM modules is lower. The increase in velocity or packing ratio of PCM, results in a higher gain as the Reynold's number is augmented. Finally, the optimum selection of PCM temperature can increase the system gains significantly. It must be ensured that PCM is fully molten by the end of the charging period.

Phase change materials have been dominating different markets other than domestic heating. As it modulates temperature around the melt temperature, they have been included in the envelope [23–27]. It also suppresses losses from thermal bridging as the indoor temperature is more uniform. It has been also included in the power generation owing to the large latent heat [28]. It has been also used in space cooling [29–31] and air conditioning [32, 33]. When PCM melt temperature is carefully selected, heating/cooling loads are reduced [34, 35]. The high storage capacity of phase change which accumulates during the day is discharged during the night.

High storage capacity of latent storage has motivated researchers to further exploit this capability. This was fulfilled by a multi-PCM configuration. This configuration can divide the whole operating range to narrow bands. This will increase

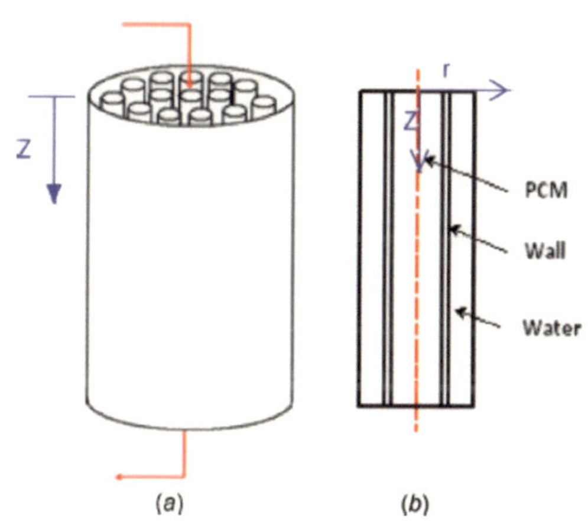

Figure 5.
Hybrid storage studied by Teamah et al. [21]. (a) whole problem domain, and (b) considered element.

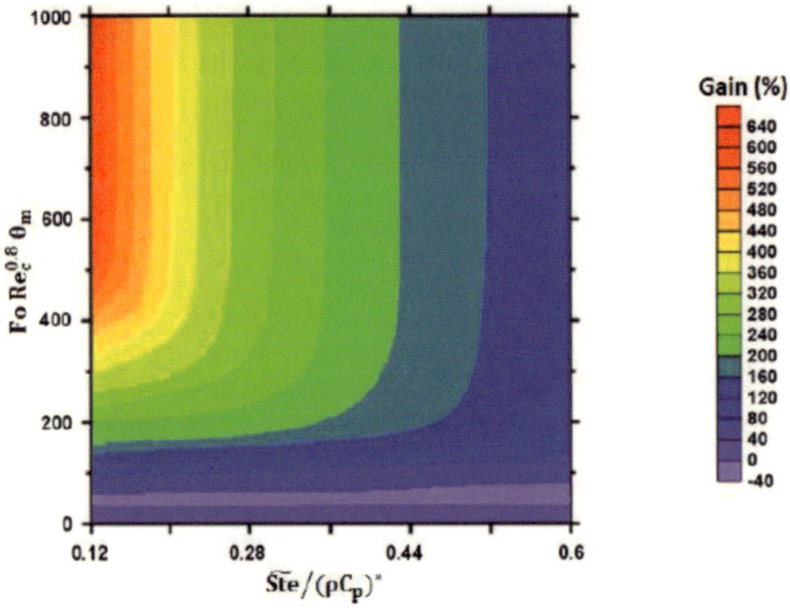

Figure 6.
Nondimensional map that correlates gain to various operating parameters (Teamah et al. [1, 21]).

the energy storage capacity. The investigation of multiple PCMs has emerged since the 1980s. There are different arrangements for such a system (**Figure 7**) [36]. It can be either in parallel or in series. However, the series arrangement is the one that harness the potential of temperature range division and higher storage potential. Michels and Pitz-Paal [37] performed a detailed comparison for single PCM system, and multi-PCM systems. They also compared it to water-based system. They considered the application of PCMs in concentrated solar power applications where the melting point is high. The multiple PCM configuration is shown in **Figure 8** [36]. They quantified that the energy storage potential is 74% higher in the multiple PCM case. The multiple PCM configuration accelerates charging and discharging of system and promotes exergy of the system [38]. A study was done on a heat exchanger with multiple PCM [39]. The PCM is in the tubes and water flows in the shell. They highlighted that the careful choice of PCMs melt temperatures is crucial. The majority of the PCM needs to be molten by the end of the charging period. If this is not guaranteed, the potential of multi-PCM decreases drastically.

Teamah et al. [40] investigated the incorporation of multiple fatty acids in cascaded tanks (**Figure 9**). They found that the multiple PCM configuration can

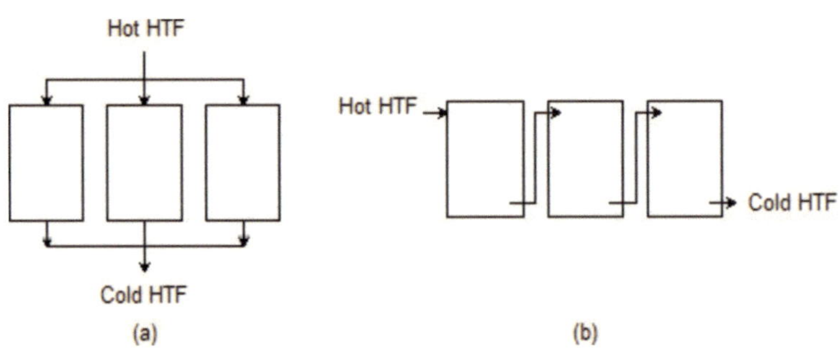

Figure 7.
Multiple phase change material systems arrangement [36]. (a) parallel configuration, and (b) series configuration.

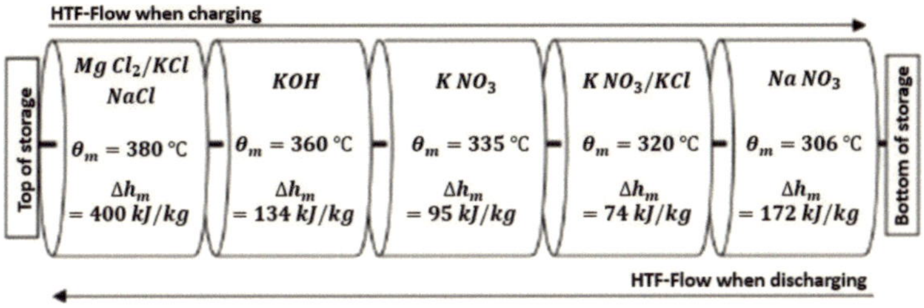

Figure 8.
Multiple PCM arrangement in Michels and Pitz-Paal [36] work.

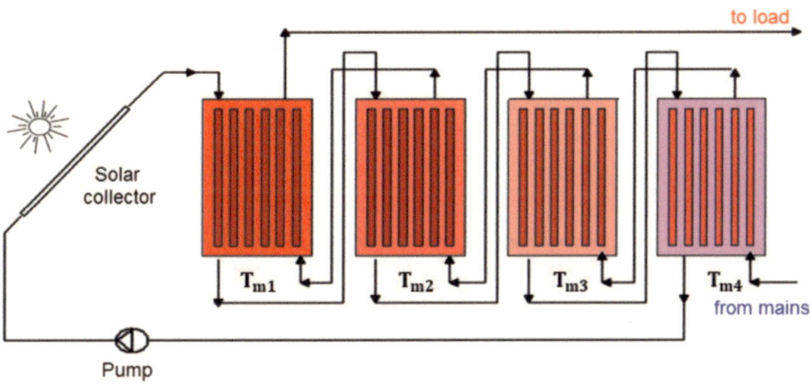

Figure 9.
Multi-tank configuration investigated by Teamah et al. [40].

increase the energy storage three times compared to the water-based system. Comparison between direct and indirect system has been performed. When water is admitted indirectly to the tank, the charging occurs through coil heat exchangers. Direct system is proven to be more efficient than indirect one. The energy stored in the direct system is also higher than the indirect system.

3. Conclusion

The introductory chapter of the book has presented the reader with basic knowledge needed to be an expert in the thermal energy storage field. It focused on the comparison between sensible and latent storage. The sensible storage is useful if the operating temperature range is higher. Latent storage system presents a great opportunity for storing heat in the narrow operating ranges. Phase change materials are used in variety of applications in the residential and commercial sector. It can stabilize the operation of different systems. Lastly multiple phase change materials can be deployed to magnify the energy storage potential.

Author details

Hebatallah Teamah
Department of Building Science, Office of Applied Research, Innovation and Entrepreneurship, Algonquin College Center for Construction Excellence, Ottawa, Canada

*Address all correspondence to: teamahh@algonquincollege.com

References

[1] H. M. Teamah, M. F. Lightstone, J. S. Cotton, Numerical Investigation and Nondimensional Analysis of the Dynamic Performance of a Thermal Energy Storage System Containing Phase Change Materials and Liquid Water. ASME. I. Sol. Energy Eng. 139 (2017), 021004-021004-14.

[2] A. Abhat, Low temperature latent thermal energy storage system: heat storage materials, Solar Energy 30 (1983), 313-332.

[3] H. Mhiri, A. Jemni, H. Sammouda, Numerical and experimental investigations of melting process of composite material (nanoPCM/carbon foam) used for thermal energy storage, Journal of Energy Storage, 29 (2020), 101167.

[4] Ebrahimi, A. Dadvand, Simulation of melting of a nano-enhanced phase change material (NePCM) in a square cavity with two heat source–sink pairs, Alexandria Engineering Journal 54 (2015), 1003-1017.

[5] W. G. Alshaer, S. A. Nada, M. A. Rady, CedricLe Bot, ElenaPalomo Del Barrio, Numerical investigations of using carbon foam/PCM/Nano carbon tubes composites in thermal management of electronic equipment, Energy Conversion and Management 89 (2015), 873-884.

[6] W. G. Alshaer, M. A. Rady, S. A. Nada *et al,* An experimental investigation of using carbon foam–PCM–MWCNTs composite materials for thermal management of electronic devices under pulsed power modes, Heat Mass Transfer 53 (2017), 569-579.

[7] W. G. Alshaer, S. A. Nada, M. A. Rady, CedricLe Bot, ElenaPalomo Del Barrio, Thermal management of electronic devices using carbon foam and PCM/nano-composite, International Journal of Thermal Sciences 89 (2015), 79-86.

[8] B. Nie, A. Palacios, B. Zou, J. Liu, T. Zhang, Y. Li, Review on phase change materials for cold thermal energy storage applications, Renewable and Sustainable Energy Reviews 134 (2020), 110340.

[9] C. Liu, C. Luo, T. Xu, P. Lv, Z. Rao, Experimental study on the thermal performance of capric acid-myristyl alcohol/expanded perlite composite phase change materials for thermal energy storage, Solar Energy 191 (2019), 585-595.

[10] M. Li, Z. Wu, A review of intercalation composite phase change material: preparation, structure, and properties, Renew Sustain Energy Rev 16 (2012), 2094-2101.

[11] Y. Dutil, D. R. Rousse, N. B. Salah, S. Lassue, L. Zalewski, A review on phase-change materials: Mathematical modeling and simulations, Renewable and Sustainable Energy Reviews, 15 (2011), 112-130.

[12] C. Beust, E. Franquet, J. Bédécarrats, P. Garcia, A numerical investigation of some key factors for the simulation of convection-dominated melting International Journal of Thermal Sciences 161 (2021), 106687.

[13] P. Chakraborty, Enthalpy porosity model for melting and solidification of pure-substances with large difference in phase specific heats, International Communications in Heat and Mass Transfer 81 (2017), 183-189.

[14] J. Dauvergne, Á Serrano, E. Palomo, D. Barrio, Fast estimation of the enthalpy–temperature function of Phase Change Materials, Experimental Thermal and Fluid Science 122 (2021), 110317.

[15] Z. Younsi, H. Naji, A numerical investigation of melting phase change process via the enthalpy-porosity approach: Application to hydrated salts, International Communications in Heat and Mass Transfer 86 (2017), 12-24.

[16] Y. Wei-An, Y. Hong-Xiao, Z. Chong, H. Xiao-Fei, Precise integration boundary element method for solving dual phase change problems based on the effective heat capacity model, Engineering Analysis with Boundary Elements 108 (2019), 411-421.

[17] I. Medved, A. Trník, L. Vozár, Modeling of heat capacity peaks and enthalpy jumps of phase-change materials used for thermal energy storage, International Journal of Heat and Mass Transfer 107 (2017), 123-132.

[18] Y. Khattari, T. Rhafiki, N. Choab, T. Kousksou, M. Alaphilippe, Y. Zeraouli, Apparent heat capacity method to investigate heat transfer in a composite phase change material, Journal of Energy Storage 28 (2020), 101239.

[19] T. Davin, A. Guillet., Supercooling of phase change: A new modeling formulation using apparent specific heat capacity," International Journal of Thermal Sciences, 147 (2020), 106121.

[20] Y. Wei-An, Y. Hong-Xiao, Z. Chong, H. Xiao-Fei, Precise integration boundary element method for solving dual phase change problems based on the effective heat capacity model, Engineering Analysis with Boundary Elements, 108 (2019), 411-421.

[21] H. M. Teamah, M.F. Lightstone, Numerical study of the electrical load shift capability of a ground source heat pump system with phase change thermal storage Energy and Buildings 199 (2019), 235-246.

[22] H. M. Teamah, Comprehensive review of the application of phase change materials in residential heating applications, Alexandria Engineering Journal 60 (2021), 3829-3843.

[23] P. A. Mirzaei, F. Haghighat, Modelling of phase change materials for applications in whole building simulation, Renew Sustain Energy Rev 16 (2012), 5355-5362.

[24] V. V. Tyagi, S. C. Kaushik, S. K. Tyagi, T. Akiyama, Development of phase change materials based microencapsulated technology for buildings: a review, Renew Sustain Energy Rev 15 (2011), 1373-1391.

[25] Z. Liu, Z. Yu, T. Yang, D. Qin, S. Li, G. Zhang, F. Haghighat, M. Joybari, A review on macro-encapsulated phase change material for building envelope applications, Building and Environment 144 (2018), 281-294.

[26] V. Sharma, A. Rai, Performance assessment of residential building envelopes enhanced with phase change materials, Energy and Buildings 208 (2020), 109664.

[27] Y. Lakhdari, S. Chikh, and A. Campo, Analysis of the thermal response of a dual phase change material embedded in a multi-layered building envelope, Applied Thermal Engineering 179 (2020), 115502.

[28] A. Gil, M. Medrano, I. Martorell, A. Lazaro, P. Dolado, B. Zalba, et al., State of the art on high temperature thermal energy storage for power generation. Part I – concepts, materials and modulization, Renew Sustain Energy Rev 14 (2010), 31-55.

[29] A. Waqas, Z. U. Din, Phase change material (PCM) storage for free cooling of buildings – a review, Renew Sustain Energy Rev 18 (2013), 607-625.

[30] G. Li, Y. Hwang, R. Radermacher, Review of cold storage materials for air conditioning application, Int J Refrig 35 (2012), 2053-2077.

[31] M. Rastogi, A. Chauhan, R. Vaish, A. Kishan, Selection and performance assessment of phase change materials for heating, ventilation and air conditioning applications, Energy Convers Manag 89 (2015), 260-269.

[32] H. Xu, J. Sze, A. Romgnoli, X. Py, Selection of Phase Change Material for Thermal Energy Storage in Solar Air Conditioning Systems, Energy Procedia 105 (2017), 4281-4288.

[33] B. Nie, Z. Du, B. Zou, Y. Li, Y. Ding, Performance enhancement of a phase-change-material based thermal energy storage device for air-conditioning applications, Energy and Buildings 214 (2020), 109895.

[34] S. A. Nada, W. G. Alshaer, R. M. Saleh, Effect of phase change material plates' arrangements on charging and discharging of energy storage in building air free cooling, Energy Storage 2020, 2 (4), e142.

[35] S. A. Nada, W. G. Alshaer, R.M. Saleh, Experimental investigation of PCM transient performance in free cooling of the fresh air of air conditioning systems, Journal of Building Engineering 29 (2020), 101153

[36] H. E. Qarnia, Numerical analysis of a coupled solar collector latent heat storage unit using various phase change materials for heating the water, Energy Convers Manag 50 (2009), 247-254.

[37] H. Michels, R. Pitz-Paal, Cascaded latent heat storage for parabolic trough solar power plants, Solar Energy, 81(2007), 829-837.

[38] T. Watanabe, A. Kanzawa, A., Second law optimization of a latent heat storage system with PCMs having different melting points, Heat Recovery Systems & CHP, Vol. 15 (1995), 641-653.

[39] T. Kousksou, P. Bruel, G. Cherreau, V. Leoussoff, T. El Rhafiki, PCM storage for solar DHW: From an unfulfilled promise to a real benefit", Solar Energy 85(2011), 2033-2040.

[40] H. M. Teamah, M. F. Lightstone, J. S. Cotton, Potential of cascaded phase change materials in enhancing the performance of solar domestic hot water systems, Solar Energy 159 (2018) 519-530.

Chapter 3

Paraffin Wax-Based Thermal Composites

Gulfam Raza, Saqib Iqbal and Abdul Samad Farooq

Abstract

Paraffin waxes are organic phase change materials possessing a great potential to store and release thermal energy. The reversible solid–liquid phase change phenomenon is the under-lying mechanism enabling the paraffin waxes as robust thermal reservoirs based on inherently high latent heat (i.e., ~200–250 J/g). However, the main drawback of paraffin waxes is their inability to expedite the phase change process owing to low thermal conductivity (i.e., ~0.19–0.35 $Wm^{-1} K^{-1}$). This drawback has long been documented as a technological challenge of paraffin waxes especially for temperature-control applications where faster thermal storage/release is necessitated, encompassing thermal management of batteries, thermoelectric modules and photovoltaic panels. Besides, sustaining the solid-like form of paraffin waxes (shape-stability) is also recommended to avoid the liquid drainage threats for crucial applications, like thermal management of buildings and fabrics. These objectives can be met by developing the paraffin wax-based thermal composites (PWTCs) with help of various thermal reinforcements. However, PWTCs also encounter severe challenges, probably due to lack of design standards. This chapter attempts presenting the recent advances and major bottlenecks of PWTCs, as well as proposing the design standards for optimal PWTCs. Also, the fundamental classification of phase change phenomenon, paraffin waxes and potential thermal reinforcements is thoroughly included.

Keywords: Paraffin wax, thermal energy storage, Phase change materials, Latent heat, thermal reinforcements

1. Introduction

Paraffin waxes are defined as the materials consisting of saturated carbon-hydrogen chains integrated with branched, straight and ring-like (aromatics) structures [1] that are relatively complex by nature. This chemical configuration endows the amorphous characteristics and inertness to paraffin waxes, resulting in inactive functional groups where the external chemical reactions become impossible. Based on this stance, paraffin waxes are supposed to be the green thermal reservoirs, lying within the sustainable targets of the current era. Therefore, the applications of paraffin waxes ranging from biomedical [2] to thermal storage/release [3–5] are declared relatively safe and environmental-friendly.

In thermal storage/release applications, phase change of the paraffin waxes is the backbone that governs the under-lying mechanism depending on thermal excitation or de-excitation driven by the heat source. Thermal excitation is the phase change process of melting during which thermal energy is absorbed, while thermal de-excitation is the phase change process of solidification during which thermal energy is released. These both processes build up the reversible functionality involving the sensible heat and latent heat storage/release. In case of paraffin waxes (or generally for PCMs), the sensible heat is counted before the phase change process, while latent heat is considered during the phase change process. By definition, phase change [6] process refers to either structural change or state change, as shown in **Figure 1**. In structural change, a single phase of paraffin wax undergoes thermal excitation that brings about the conversion of the internal structures and it is called the solid–solid phase change. Whilst, in the state change, two phases of the paraffin wax undergoes thermal excitation, resulting in conversion of solid phase into liquid phase at the melting temperature, which is known as solid–liquid phase change. The occurrence of either kind of phase change is dependent on the melting temperatures of the paraffin waxes. The paraffin waxes with low-melting temperature such as <40°C (or also called soft paraffin waxes) do not demonstrate structural change, so they only undergo state change. While, the paraffin waxes with high-melting temperature (or also called hard paraffin waxes) provide both structural change and state change.

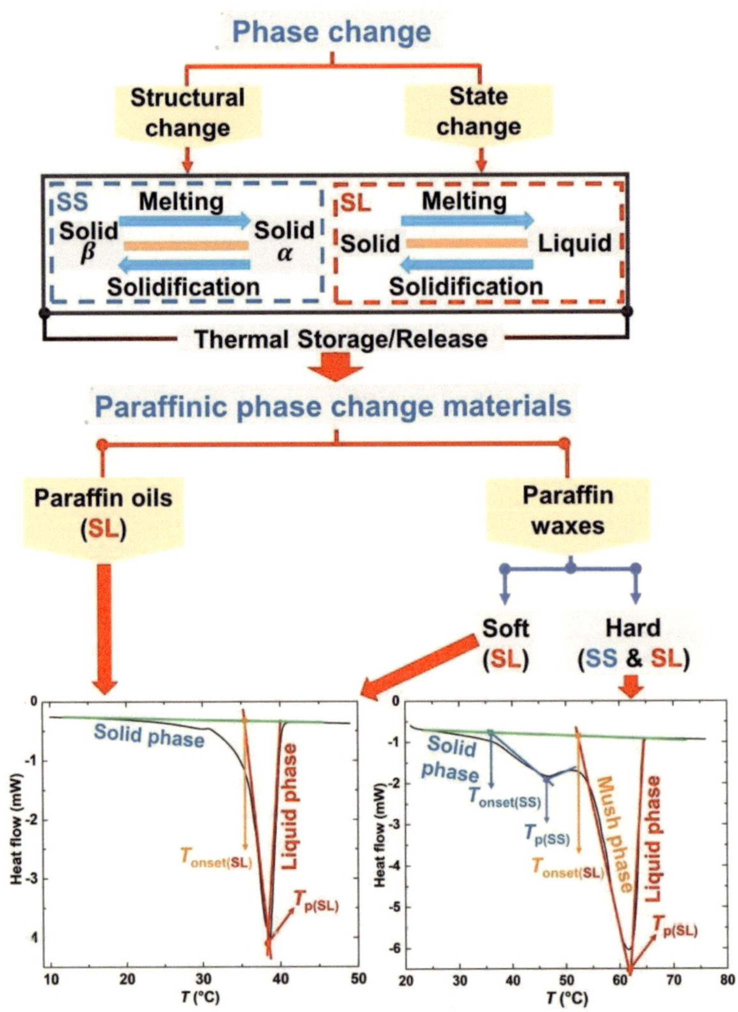

Figure 1.
Classification of phase change and paraffin wax-based phase change materials. A few contents of this figure are partially restructured from references [6, 7].

In between structural change and state change, another phase consisting of solid flakes and liquid sites ensues which is called mush phase. Normally, the temperature range of the mush phase lies closer to the onset temperature of the liquid phase. In brief, all these phases are responsible for the total latent heat, while before and after these phases, total sensible heat is considered, depending on the melting temperatures of the involved paraffin waxes. Therefore, it is emphasized that the total thermal energy of the paraffin waxes should carefully include thorough observation of the onset and endset temperatures which can be determined either by DSC curves (**Figure 1**) or by the transient heat diffusion process (temperature versus time analysis). With help of transient heat diffusion process, the thermal storage/release performance of paraffin waxes can precisely be assessed by looking into the time consumed during the melting (thermal excitation) and solidification (thermal de-excitation) processes, which is called charging time and discharging time, respectively. It is generally deducted that, the charging/discharging time becomes short, if the heat diffusion occurs fast, and vice versa. The heat diffusion is further related to the intrinsic property of materials known as thermal conductivity, which is defined as the ability of material allowing the fast or slow transfer of heat. In short, with high thermal conductivity, the charging/discharging time are reduced on behalf of faster heat transfer. However, heat transfer in paraffin waxes is sluggish, and their charging/discharging time is sufficiently long based on the low thermal conductivity. The possible reason for the low thermal conductivity of paraffin waxes is their amorphous nature wherein tightly-packed and inter-connected thermal paths are unavailable. However, a great care is essential because every application may not need fast heat transfer rates, for example, thermal management of buildings where the objective is to keep the heat either outside the buildings in summer or inside the buildings in winter, which is possible only if paraffin wax serves as a thermal insulator necessitating slow heat transfer rates, but the challenge of liquid drainage needs to be simultaneously addressed. Therefore, depending on the target applications, enhancing the thermal conductivity of paraffin waxes is of great practical interest so that the charging/discharging can be reduced. For example, the long charging/discharging time of paraffin waxes is the major bottleneck that can potentially hampers their functionality for thermal management of batteries where the heat generation rate is prone to high and the objective is to dissipate the heat so that hot-spots can be avoided. Keeping different practical scenarios ahead, thermal reinforcements are essential to improve the thermo-physical bottlenecks of the base materials (paraffin waxes), helping create paraffin wax-based thermal composites (PWTCs) which are discussed henceforth.

2. Potential thermal reinforcements

Thermal reinforcements are defined as the external additives being mingled with paraffin waxes to fabricate PWTCs. Thermal reinforcements need to possess certain traits encompassing high/low thermal conductivity, good mechanical strength, porous surface geometry, adaptability and the uniform mingling-compatibility with the paraffin waxes. Depending on the chemistry of thermal reinforcements, they are majorly categorized into inorganic sorbents and organic encapsulants, as depicted in **Figure 2**. Inorganic sorbents are defined as thermal reinforcements providing the physical scaffolds wherein paraffin wax is infused through capillary action. These scaffolds can also be said as oleophilic (be noted all paraffinic PCMs have oiliness though the percentage may differ) that all paraffin waxes based on the wetting characteristics induced by micro/nano-porous structures. Inorganic sorbents are further classified into carbonic, silicate and metallic

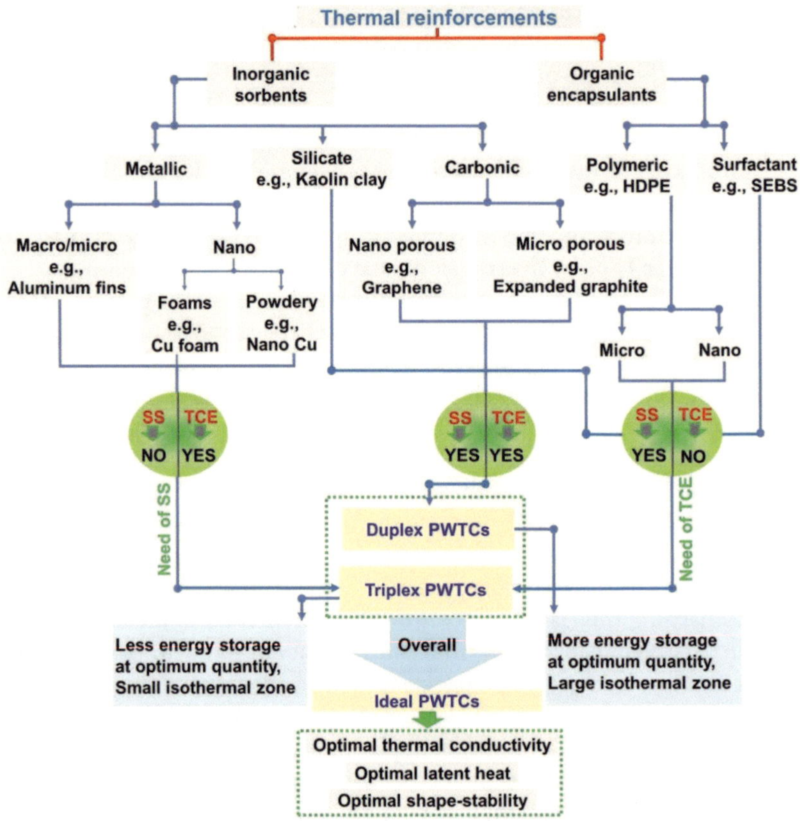

Figure 2.
Classification of thermal reinforcements and their role as thermal conductivity enhancer (TCE) or shape-stabilizer (SS). A few contents of this figure are partially restructured from references [6].

thermal reinforcements. Organic encapsulants are defined as thermal reinforcements that build a shell (a boundary encircling the entity) wherein paraffin wax resides as the core material. The next section sheds a light on the preparation methods, morphologies and thermo-physical properties of PWTCs made of these thermal reinforcements.

In the light of application-dependent scenarios, there are two functions that PWTCs should endow to the paraffin wax: temperature-control capacity and shape-stability, but the existence of both functions may not be found in single PWTC. Temperature-control capacity is a coupled function of latent heat and thermal conductivity of PWTCs, imparting a suitable isothermal zone that is highly desired in thermal management applications. Thermal reinforcements that help enhances thermal conductivity of the paraffin waxes are called thermal conductivity enhancers (TCEs). As for the shape-stability of PWTCs, it is defined as the state in which the liquid drainage of the melted paraffin wax is minimized or eventually stopped, and such thermal reinforcements are termed as shape-stabilizers (SSs). In brief, thermal reinforcements can potentially act as the TCSs, SSs or both, and by incorporating the most appropriate thermal reinforcements; duplex or triplex PWTCs can be fabricated.

2.1 Inorganic sorbents

Inorganic sorbents and the corresponding PWTCs are discussed in the coming subsections.

2.1.1 Carbonic thermal reinforcements

Carbonic thermal reinforcements are documented as inherently high-thermal conductivity materials based on well-connected thermal networks. They majorly include graphite powder (PG), expanded graphite (EG, which is obtained by heat-treating the expandable graphite at high temperature of 800–900°C for a few seconds), single-wall carbon nano-tubes (SWCNTs), multi-wall carbon nano-tubes (MWCNTs), and graphene nano-platelets (GNPs). The salient properties of the carbonic thermal reinforcement include the appropriate infusion-compatibility and unique morphology consisting of micro/nano-porous structures, as shown in **Figure 3**, rendering them both TCEs and SSs.

However, the disadvantageous features of carbonic thermal reinforcements also exist, for example, anisotropic thermal conductivity of graphite, reduced thermal conductivity of graphene when it is mixed, and segregation of carbon nano-tubes. In addition, duplex and triplex PWTCs can be synthesized with them via dry or wet-physical methods. For example, a triplex PWTC has been fabricated by dry-physical method, employing 80% of paraffin wax (melting temperature of 48–50°C, thermal conductivity of, and latent heat of 207 jg^{-1}), 20% of EG and 5% of GNPs. In addition to EG, the incorporation of GNPs has further introduced effective results with enhanced thermal conductivity (around 5.9 $Wm^{-1} K^{-1}$ at compress density of 505 kgm^{-3}) and latent heat of 159 jg^{-1} [8], which is thus deemed to be feasible because the mass percent of all ingredients is somehow optimal. Nonetheless, graphene (also the graphene oxide) has been reported to have very high thermal conductivity (5000 $Wm^{-1} K^{-1}$), but EG/PWTC/GNP could not correspondingly achieve that high thermal conductivity, implying a surprising phenomenon. The reason is attributed to the design challenge of GNPs, i.e., when GNPs are joined together to form a compound, the phase segregation effect appears causing to cleave the internal thermal networks due to which the thermal conductivity of compound GNPs is much reduced [10]. Another case study of triplex PWTC [11] has been presented, consisting of 50% paraffin wax (melting temperature of 60–61°C, thermal conductivity of 0.26 $Wm^{-1} K^{-1}$, and latent heat of 223 jg^{-1}), 10% of GP and

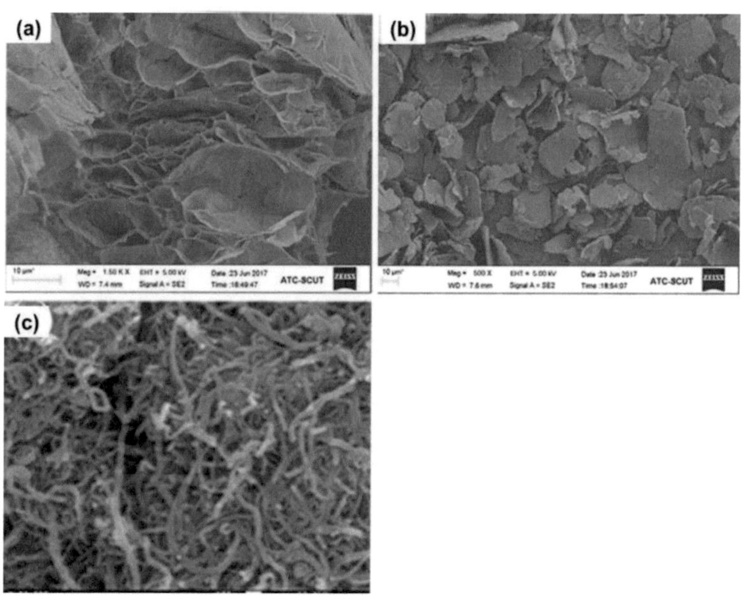

Figure 3.
Micro/nano-porous morphology of (a) EG [8], (b) GNPs [8] and (c) CNTs [9].

40% of expanded perlite (EP). As-prepared GP/EP/PWTC has latent heat of 111.4 jg-1 and thermal conductivity of 1.34 $Wm^{-1} K^{-1}$. The question arises on the latent heat which reduces from 223 jg^{-1} to 111.4 jg^{-1} which is mainly because of 50% of EP. In the meanwhile, the EP/PWTC neither achieves high thermal conductivity nor shape-stability, while upon incorporation of 10% GP, high thermal conductivity as well as effective shape-stability has been attained, as shown in **Figure 4** [11].

As a conclusive viewpoint, the need of thermal reinforcements should be well-understood together with avoiding the unimportant thermal reinforcements while synthesizing PWTCs, which otherwise can definitely lead to severe shortcomings particularly for energy storage applications where high thermal storage capability is preferred. Therefore, instead of such triplex PWTCs (e.g., EP/GP/paraffin wax), duplex PWTCs (e.g., GP/paraffin wax) may result in more viable outcomes.

The anisotropic thermal conductivity of EG has a great influence on thermo-physical properties of the paraffin wax. The layered sheet/wall-like structure of EG is thus accountable for this effect, inhibiting the heat transfer when it is perpendicular to the graphite layered-structure, while expediting the heat transfer when it is parallel. Such anisotropic effects of has been studied recently for EG alone and EG/PWTCs, demonstrating the much higher thermal conductivity in parallel direction than that of normal direction depending on various temperatures (**Figure 5**) [12].

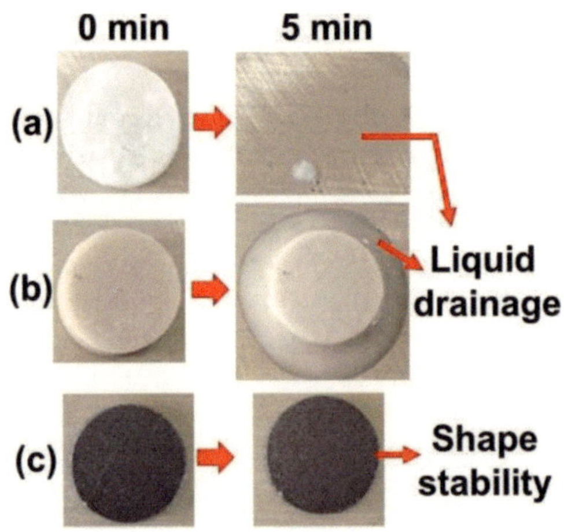

Figure 4.
Shape-instability of (a) paraffin wax and (b) EP/PWTC, and (c) shape-stability of EP/GP/PWTC [11].

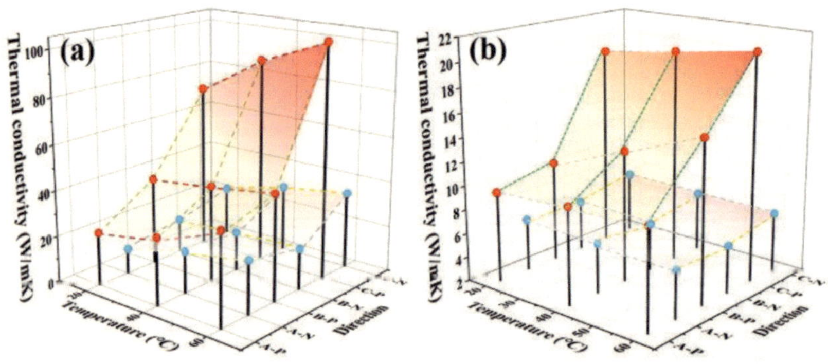

Figure 5.
Directional thermal conductivity of (a) EG and (b) EG/PWTC. P: Parallel direction, N: Normal direction [12].

These results imply that the directional effects should also be considered while implementing the EG/PWTCs in applications. In addition, particle size effect of EG is also important. It has been investigated recently that the large-sized and small-sized particles of EG have thermal conductivities of 0.5 $Wm^{-1} K^{-1}$ and 3.23 $Wm^{-1} K^{-1}$, respectively, which are capable of enhancing thermal conductivity of PWTCs in the same order [13]. In general, thermal conductivity of PWTCs increase with increasing the percent contents of thermal reinforcement. However, it is achieved on the penalty of latent heat reduction [13]. This is the reason that PWTCs need to be optimized by keeping the design parameters ahead, so that the ideal PWTCs can be created as claimed above in **Figure 2**.

Thermal stability of paraffin waxes is defined as the maximum temperature limit after which thermal decomposition begins and it is normally 150–170°C [9]. Evaluated with help of thermogravimetric analysis (in which the specimen material is evaporated at high temperature while simultaneously measuring the mass loss indicating the thermal decomposition), carbonic/PWTCs have been reported with enhanced thermal stability (around 190–200°C) [13] which is indeed attributed to the heat-withstanding strength of thermal reinforcements. The enhanced thermal stability also indicates the successful infusion of paraffin waxes into thermal reinforcements which then impart a kind of thermal protection during thermal processes.

2.1.2 Silicate-based thermal reinforcements

The silicate-based thermal reinforcements are regarded as low thermal conductivity materials, but their micro/nano-porous morphology (**Figure 6**) and consequential specific surface area allows them to impregnate with the paraffin waxes. More precisely, they come in class of clay-like materials and commonly termed as the clay minerals, for examples, expanded perlite (EP), kaolin clay, diatomite, palygorskite or attapulgite and vermiculate, etc., [14–16].

The silicate-based thermal reinforcements are available in abundance at cheap and economical rates, as well as they are notably non-toxic and have a good compatibility with the paraffin waxes based on which shape-stabilized PWTCs can be achieved. However, a large percent threshold of silicate-based thermal reinforcements is normally required to functionalize them as SS for PWTCs. In addition, for high thermal response of PWTCs with silicate-based thermal reinforcements, TCE in sufficient percent may be required; hence, the triplex PWTC cannot be avoided in that case, for example, carbonic TCEs have been incorporated [17]. In the meanwhile, the main point of focus, which is of great practical significance, is the consideration of applications of silicate/PWTCs. Normally, the most compatible features of silicate-based thermal reinforcements have been seen appropriate for the building materials, such as cement, gypsum and sand, etc., implying that

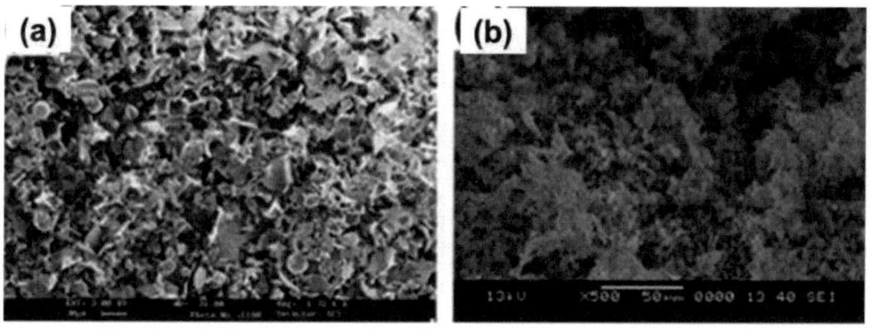

Figure 6.
Micro/nano-porous morphology of (a) EP [14] and (b) diatomite [15].

thermal management of buildings can be done if PWTCs. In such applications, very high latent heat enabled by large percent contents of paraffin waxes is completely undesired. For example, 20% of paraffin wax/expanded perlite is declared to be the optimum in the main skeleton of cement mortar [17]. The reason is the non-stiffness of paraffin waxes, leading to decrease the flexural and compressive strength of building materials at percent contents. Therefore, the quantitative optimization of silicate/PWTCs is highly crucial. Based on this design principle, low latent heat of paraffin wax/diatomite/gypsum triplex composite (around 45 J/g) has been found quite reasonable [18]. Nonetheless, enhancing the thermal response of silicate/PWTCs for building applications is debatable on the standpoint: buildings need to reserve the heat inside the room in winter or outside of it in summer, but high thermal conductivity of silicate/PWTCs is expected to boost up the heat transfer rate which is almost similar to the cement-based walls. For example, a case-study is depicted in **Figure 7** [19]. In heating mode (**Figure 7a**), the paraffin wax/diatomite thermal composite a larger isothermal zone (blue curve in **Figure 7a**) compared with that of paraffin wax/diatomite/CNT (red curve in **Figure 7a**), while in cooling mode (**Figure 7b**), the opposite trend holds true. This is ascribed to the high thermal conductivity of PWTCs achieved on behalf of CNTs. Suppose that this PWTC is applied in building walls in hot countries where the average temperature in summer is higher than 45; the high thermal conductivity of PWTC is deemed to allow heat transfer at fast rates, meaning that the time consumed in saturating the PWTC is less. With this trend, the overall time to keep the thermal management of buildings, both in heating and cooling modes, is expected to be decreased. In simple words, the isothermal zone should be long-lasting so that the more time can be ensured for thermal management. Therefore, there is a great need to decide whether silicate/PWTCs should have high or low thermal conductivity.

2.1.3 Metallic thermal reinforcements

Metallic thermal reinforcements are regarded as highly thermal conductive materials existing in three scaffolds, namely: fins that are extruded plate/tube-like thin structures, foams consisting of wire/fiber-based network with varying degree of number of pores, and powders that are composed of micro/nano-particles [20]. The micro/nano-porous scaffolds of metallic thermal reinforcements serve as the confinement sites wherein paraffin waxes reside, as demonstrated the surface morphologies in **Figure 8**.

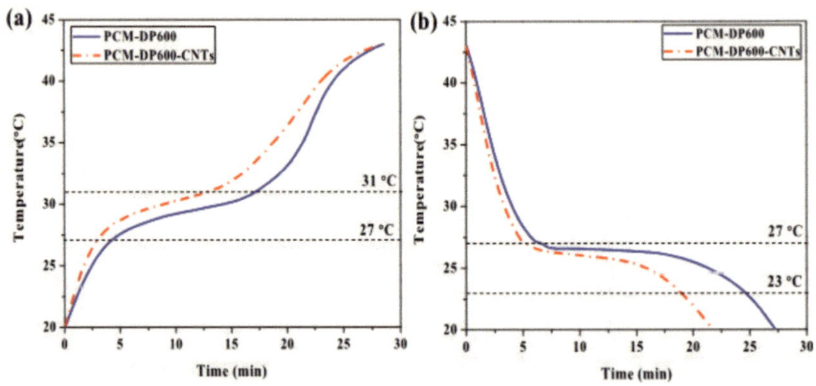

Figure 7.
(a) Heating mode, and (b) cooling mode of PWTCs via transient thermal analysis [19].

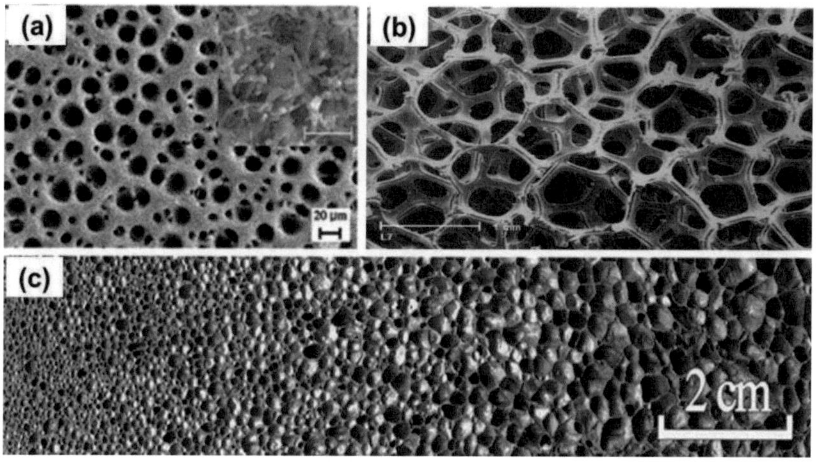

Figure 8.
Micro/nano-porous scaffolds of (a) copper foam [21], (b) nickel foam [22] and (c) graded aluminum foam [23].

Metallic materials own a very high thermal conductivity owing to the freely available electronic carries and vibration-assisted modes of heat transfer. The same metallic material, for examples, copper, nickel and aluminum, can be either available in foam or powder, but their initial preparation methods differ a lot. Metallic thermal reinforcements can only act as TCEs, and shape-stability cannot be ensured since the interaction between their network and paraffin waxes is solely based on capillary forces lacking of the liquid-soaking capability.

This drawback may create additional challenges such as weakening the thermal interface between paraffin wax and metallic fibers. Nonetheless, the effective thermal conductivity of PWTCs via copper foams has been achieved very high, such as 16 $Wm^{-1} K^{-1}$ which is based on the high inherent thermal conductivity of copper foam (i.e., 400 $Wm^{-1} K^{-1}$) [24]. Overall, the work on improving the weak thermal interface is left as a research area of future.

2.2 Organic encapsulants

The organic encapsulants may be developed in two structural scales, such as nano or micro. They include polymers and surfactants that are being introduced to develop PWTCs. It is worth-noting that the surfactants are normally polymers, but precisely, they are thermo-plastic elastomers consisting of co-polymers blocks. The potential polymeric encapsulants include low-density polyethylene (LDPE), high-density polyethylene (HDPE), melamine-formaldehyde (MF), polypropylene and polyacrylate, etc., while surfactant-based encapsulants are styrene-b-(ethylene-co-butylene)-b-styrene (SEBS) tri-block copolymer, styrene–butadiene–styrene (SBS), olefin block copolymer (OBC) and polystyrene, among others. However, the main challenge with organic encapsulants is that their thermal conductivity is approximately similar to that of paraffin waxes [25]. Therefore, organic encapsulants are only SSs. The need of TCEs with encapsulated PWTCs definitely results in a triplex thermal composite, implying that the latent heat has to be further sacrificed. Such a design challenge can only be resolved by recommending the encapsulated PWTCs for those applications where thermal energy storage is moderately acquired but fast charging/discharging is not a primary objective, for example, in flame retardancy and thermal management of buildings, fabrics and food packages. The fabrication of encapsulated PWTCs is based on dry or wet-physical methods;

however, the fabrication of the encapsulants is usually enabled by the chemical reactions which are complex due to the inclusion of several preparation steps and process conditions [26]. The little variance in process conditions can result in broken capsules (shells), posing the liquid drainage threats for PWTCs. Until recently, encapsulated PWTCs have been successfully fabricated. For example, duplex encapsulated PWTCs have been presented comprising of paraffin wax (melting temperature of 52–54 C, latent heat of 182.6 Jg^{-1}) as a core and MF as shell, as shown in **Figure 9** [27]. With 10 g of paraffin wax and 61.6% encapsulation efficiency, the latent of encapsulated PWTC is 107.4 Jg^{-1}.

It is thus obvious that the encapsulation process covers the several steps at various process constraints to create the final PWTCs. Normally, the morphology is spherical and it is filled with paraffin wax, as shown in Figure, therefore encapsulation efficiency (how much core material can be surrounded and uptaken by the shell of encapsulant) is the fundamental parameter to be emphasized for encapsulated PWTCs. In another example of triplex PWTC is presented where HDPE has been employed as an encapsulant for paraffin wax (melting temperature of 54–56°C and latent heat of 199 jg^{-1}), and copper metal foam has been introduced enhancing the thermal conductivity of HDPE/paraffin wax from 0.72 Wm^{-1} K^{-1} to 2.14 Wm^{-1} K^{-1} and reducing the latent heat as minimum as 151.6 jg^{-1}. The PWTCs also

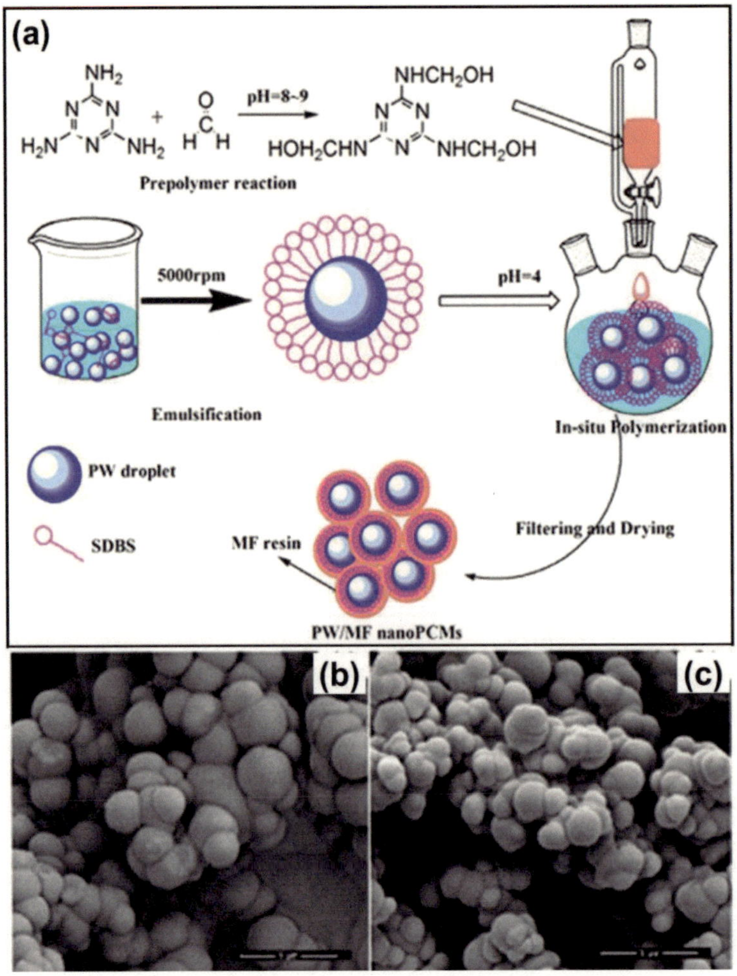

Figure 9.
(a) Fabrication of encapsulated PWTC and (b-c) the corresponding morphology [27].

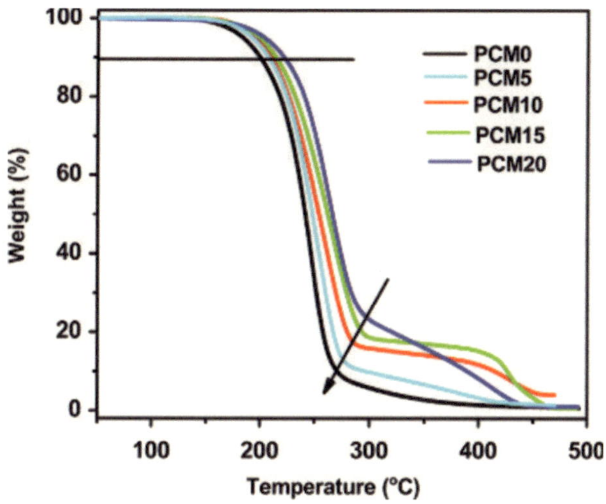

Figure 10.
Thermal stability of paraffin wax and PWTCs via TGA [30].

provided favorable shape-stability, allowing 2.3% mass loss of paraffin wax after 50 thermal cycles [28]. In a word, the stronger the encapsulant shell is, the more effective is the shape-stability. Generally, several other examples are also found where carbonic thermal reinforcements have been introduced to accelerate thermal response of encapsulated PWTCs, such as EG [25, 26].

Apart from the salient merits of providing shape-stabilization, organic encapsulants are thermally stable too [29], imparting the adequate thermal stability to PWTCs. For example, SEBS has been employed to encapsulate paraffin wax (melting temperature of 52–54°C and latent heat of 176.6 jg^{-1}) in different percent contents (5–20%), and 20% of SEBS enhances the thermal decomposition temperature of PWTC to around 221.4°C, as demonstrated in **Figure 10** [30].

3. Bottlenecks, recommendations and design standards for PWTCs

The fabrication of PWTCs is simple until investigated recently; nonetheless thermo-physical bottlenecks of the paraffin waxes still exist. For example, their inert features do not allow chemical reactions, so the utilization of chemical methods for fabrication of PWTCs can be overlooked. Thus, physical methods are only viable options for the fabrication of PWTCs. However, physical methods can also be wet or dry, for example, mixing the melted paraffin waxes with thermal reinforcement is the dry-physical method, while dissolving the paraffin wax into a solvent and then mix with thermal reinforcement is the wet-physical method or encapsulating the paraffin wax in the capsules is also the wet-physical method. Meanwhile, the shape-stabilization can be achieved via both methods, but thermal conductivity improvements have abundantly been reported based on dry-physical methods. Besides, thermal conductivity of PWTCs prepared via dry-physical methods is not sufficiently high, leaving a wide research roam and urging to put rigorous efforts into this serious matter. However, the PWTCs with increased thermal conductivity suffer from another challenge which consists of a short and uprising isothermal zone, indicating the reduced duration of temperature-control capacity. The uprising isothermal zone dictates that the temperature-control capacity of PWTC is not perfectly constant, but nearly constant and fast. A case-study describing this standpoint can be seen above in **Figure 7**.

Thus, either maximum or minimum, threshold limit of the thermal reinforcements is very important in PWTCs, which is however not standardized yet. Thermal reinforcements always demand the equal replacement of base material in PWTCs, signifying that the equivalent reduction in latent heat storage capability [12]. In such a situation, the design parameters of PWTCs should be adjusted according to the target applications. For example, thermal management of photovoltaic panel is required to be done through PWTC. In this case, the liquid drainage of the PWTC may not be considered a design parameter of the primary importance because the mechanical enclosure can assist in controlling the liquid drainage, meaning that the form-stability of PWTC can be neglected. However, the heat accumulation at the interface of PV and the hot-spots in the PWTC body are altogether supposed to be the primary design parameters. Therefore, thermal reinforcements for such applications should solely be the TCEs that can help fabricate the duplex PWTCs. Counter-institutively, Another case can be discussed regarding the thermal management of buildings where PWTCs should capable of ensuring prolonged temperature-control capacity as well as adequate shape-stability. So, thermal reinforcements for such applications should solely be the SSs, but the SSs needs to conform the property prerequisites of the building ingredients like cement, sand, and clay, etc., which altogether leads to the triplex PWTCs. In between these two scenarios, special applications such as thermal management of satellites, robots, and astronauts can be put forth, where the total quantity of PWTC emerges is another design parameter together with consideration of temperature-control capacity and shape-stabilization.

PWTCs can also suffer from the challenge of saturation energy storage limit which is dependent on the total thermal energy storage (sensible heat + latent heat)

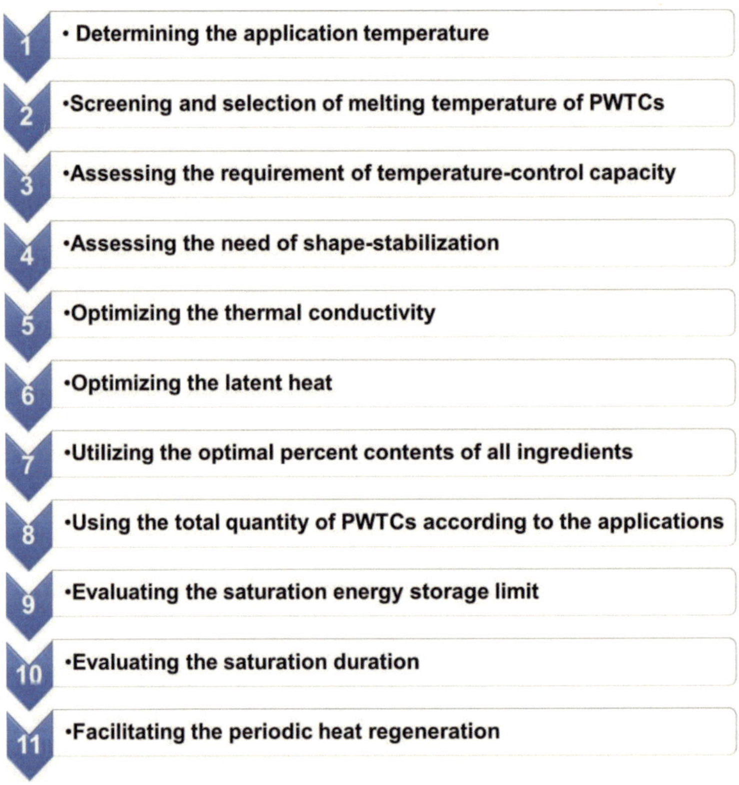

Figure 11.
Proposal of design standards for PWTCs.

of paraffin wax. This challenge is expected to emerge in all kinds of applications. Therefore, it is strongly recommended to evaluate the total saturation duration that can anticipate saturation energy storage limit. Accordingly, the periodic heat regeneration of PWTCs, i.e., extracting the stored thermal energy, becomes essential to keep the whole thermal management system fresh for the coming cycles. The possible methods for periodic heat regeneration encompass the natural convection or forced convection achieved either via air or water for which deep design considerations of the whole thermal management systems are required.

Among thermal reinforcements, EG can act both as TCE and SS [29, 31], pointing out that duplex PTWCs can be fabricated for thermal management with high charging/discharging rates. Therefore, EG is declared to be the most effective carbonic thermal reinforcement.

Although the design standardization of the optimal PWTCs is complicated, the parameters as-proposed in **Figure 11** may serve as the preliminary design principles.

4. Conclusions

The research paradigm on PWTCs has recently been shifted from duplex thermal composites to triplex thermal composites, which however lacks of the design standards. It is therefore stressed that the fabrication of PWTCs should be reported conforming with the screening requisites of specific applications, because proclaim is that every PWTCs is not applicable everywhere. Shape-stability is the design standard, but it may be essential only for a few applications, most importantly including thermal management of buildings and fabrics. Instead of shape-stability, high thermal energy storage capability and high thermal conductivity are the most important parameters for thermal management of heat-emitting devices and systems where mechanical enclosures are essential, helping in controlling the liquid drainage threats. Thus, shape-stabilizers should be carefully employed. In all, attention should be paid on thermal reinforcements such that the optimal PWTCs can be created. Most importantly, the focus should be asserted on the thermal interface between paraffin waxes and thermal reinforcement which is the backbone allowing the development of well-defined thermal paths for higher thermal conductivities of PWTCs. On the whole, three main hypotheses should be carefully checked for an ideal PWTC strictly depending on the target application: (i) preservation of optimal latent heat, (ii) maintenance of the trade-off between thermal conductivity enhancement and latent heat reduction, (iii) achievement of the optimal shape-stability.

Acknowledgements

The support provided by IntechOpen is highly acknowledged.

Conflict of interest

The authors declare no conflict of interest.

Author details

Gulfam Raza[1*], Saqib Iqbal[2] and Abdul Samad Farooq[1]

1 Institute of Refrigeration and Cryogenics, Shanghai Jiao Tong University, Shanghai, China

2 School of Chemical Engineering, Zhengzhou University, Henan, China

*Address all correspondence to: gulfamrazahaidery@hotmail.com

References

[1] Vakhshouri A. R, Paraffin as Phase Change Material. In: Soliman F.S, editor. Paraffin-an Overview. IntechOpen, 2019; DOI: 10.5772/intechopen.90487.

[2] Donczoa B, Guttman A, Biomedical analysis of formalin-fixed, paraffin-embedded tissue samples: The Holy Grail for molecular diagnostics. Journal of Pharmaceutical and Biomedical Analysis. 2018; 155: 125-134.

[3] Tariq L. S, Ali H.M, Akram M.A, Janjua M.M, Ahmadlouydarab M, Nanoparticles enhanced phase change materials (NePCMs)-A recent review. Applied Thermal Engineering. 2020; 176: 115305.

[4] Zhang P, Xiao X, Ma Z.W, A review of the composite phase change materials: Fabrication, characterization, mathematical modeling and application to performance enhancement. Applied Energy. 2016;165: 472-510.

[5] Gulfam R, Zhu Z, Xu L, Cheema I.I, Sheng P, Zhao G, Deng Y, Design, fabrication and numerical analysis of compact thermal management system integrated with composite phase change material and thermal bridge, Energy Conversion and Management. 2018; 156: 25-33.

[6] Gulfam R, Zhang P, Meng Z, Advanced thermal systems driven by paraffin-based phase change materials–A review. Applied Energy. 2019;238: 582-611.

[7] Gulfam R, Orejon D, Choi C.H, Zhang P, Phase-change slippery liquid-infused porous surfaces with thermo-responsive wetting and shedding states. ACS Applied Materials and Interfaces. 2020;12: 34306-34316.

[8] Xu T, Li Y, Chen J, Wu H, Zhou X, Zhang Z, Improving thermal management of electronic apparatus with paraffin (PA)/expanded graphite (EG)/graphene (GN) composite material. Applied Thermal Engineering. 2018; 140: 13-22.

[9] Atinafu D.G, Yun B.Y, Wi S, Kang Y, Kim S, A comparative analysis of biochar, activated carbon, expanded graphite, and multi-walled carbon nanotubes with respect to PCM loading and energy-storage capacities. Environmental Research. 2021; 195: 110853.

[10] Yang Y, Cao J, Wei N, Meng D, Wang L, Ren G, Yan R, Zhang N, Thermal conductivity of defective graphene oxide: A molecular dynamic study. Molecules. 2019; 24: 1103.

[11] Zuo X, Zhao X, Li J, Hu Y, Yang H, Chen D, Enhanced thermal conductivity of form-stable composite phase-change materials with graphite hybridizing expanded perlite/paraffin. Solar Energy. 2020; 209: 85-95.

[12] Wang X.L, Li B, Qu Z.G, Zhang J.F, Jin Z.G, Effects of graphite micro-structure evolution on the anisotropic thermal conductivity of expanded graphite/paraffin phase change materials and their thermal energy storage performance. International Journal of Heat and Mass Transfer. 2020; 155: 119853.

[13] Zhao Y, Jin L, Zou B, Qiao G, Zhang T, Cong L, Jiang F, Li C, Huang Y, Ding Y, Expanded graphite–Paraffin composite phase change materials: Effect of particle size on the composite structure and properties. Applied Thermal Engineering. 2020; 171: 115015.

[14] Bastani, D., et al., Study of oil sorption by expanded perlite at 298.15 K. Separation and Purification Technology, 2006; 52(2): 295-300.

[15] Jeong, S.-G., et al., Optimal preparation of PCM/diatomite

composites for enhancing thermal properties. International Journal of Heat and Mass Transfer. 2013; 62: 711-717.

[16] Lv P, Liu C, Rao Z, Experiment study on the thermal properties of paraffin/kaolin thermal energy storage form-stable phase change materials. Applied Energy. 2016; 182: 475-487.

[17] Sun, D. and L. Wang, Utilization of paraffin/expanded perlite materials to improve mechanical and thermal properties of cement mortar. Construction and Building Materials 2015; 101: 791-796.

[18] Lorwanishpaisarn, N., et al., Characterization of paraffin/ultrasonic-treated diatomite for use as phase change material in thermal energy storage of buildings. Journal of Thermal Analysis and Calorimetry. 2017; 128(3): 1293-1303.

[19] Xu B, Li Z, Paraffin/diatomite/multi-wall carbon nanotubes composite phase change material tailor-made for thermal energy storage cement-based composites. Energy. 2014; 72: 371-380.

[20] Rehman T, Ali H.M, Janjua M.M, Sajjad U, Yan W.M, A critical review on heat transfer augmentation of phase change materials embedded with porous materials/foams. International Journal of Heat and Mass Transfer. 2019; 135: 649-673.

[21] Kim, J.-H., R.-H. Kim, and H.-S. Kwon, Preparation of copper foam with 3 dimensionally interconnected spherical pore network by electrodeposition. Electrochemistry Communications, 2008. 10(8): p. 1148-1151.

[22] Liu, P. and K. Liang, Preparation and corresponding structure of nickel foam. Materials science and technology, 2000. 16(5): p. 575-578.

[23] He, S.-Y., et al., Preparation of density-graded aluminum foam. Materials Science and Engineering: A, 2014. 618: p. 496-499.

[24] Xiao, X., P. Zhang, and M. Li, Effective thermal conductivity of open-cell metal foams impregnated with pure paraffin for latent heat storage. International Journal of Thermal Sciences. 2014; 81: 94-105.

[25] Wang T.H, Yang T.F, Kao C.H, Yan W.M, Ghalambaz M, Paraffin core-polymer shell micro-encapsulated phase change materials and expanded graphite particles as an enhanced energy storage medium in heat exchangers. Advanced Powder Technology. 2020;31: 2421-2429.

[26] Luo D, Xiang L, Sun X, Xie L, Zhou D, Qin S, Phase-change smart lines based on paraffin-expanded graphite/polypropylene hollow fiber membrane composite phase change materials for heat storage. Energy. 2020;197: 117252.

[27] Zhang N, Yuan Y, Synthesis and thermal properties of nanoencapsulation of paraffin as phase change material for latent heat thermal energy storage. Energy and Built Environment. 1 (2020; 1: 410-416.

[28] Chen P, Gao X, Wang Y, Xu T, Fang Y, Zhang Z, Metal foam embedded in SEBS/paraffin/HDPE form-stable PCMs for thermal energy storage. Solar Energy Materials and Solar Cells. 2016;149: 60-65.

[29] Chriaa I, Karkri M, Trigui A, Jedidi I, Abdelmouleh M, Boudaya C, The performances of expanded graphite on the phase change materials composites for thermal energy storage. Polymer. 2021;212: 123128.

[30] Zhang Q, Zhao Y, Feng J, Systematic investigation on shape stability of high-efficiency SEBS/paraffin form-stable phase change materials. Solar Energy Materials and Solar Cells. 2013;118: 54-60.

[31] Wu W, Wu W, Wang S, Form-stable and thermally induced flexible composite phase change material for thermal energy storage and thermal management applications. Applied Energy. 2019; 236: 10-21.

Chapter 4

CFD Model of Shell-and-Tube Latent Heat Thermal Storage Unit Using Paraffin as a PCM

Maher Mohammad Al-Maghalseh

Abstract

This chapter validates the capability of CFD modelling technique to accurately describe processes in the thermal storage system with the PCM. For validation purposes, CFD modelling using FLUENT ANSYS was conducted and the predicted results were compared with the experimental and numerical data from the literature. The comparison between experimental and numerical results was carried out in terms of the temperature distributions and average volume of the PCM liquid fraction. Additionally, the detailed parametric study of the storage system with the PCM was performed and results obtained were discussed with dimensional correlations for the Nusselt number being proposed to be used in the designing process. Finally, a correlation was developed to estimate the total melting time at the thermal storage system.

Keywords: CFD, PCM, thermal storage system, Nusselt number, paraffin, FlUENT/ANSYS

1. Introduction

The thermal energy storage systems can be classified into several main groups, namely thermochemical storage, sensible heat storage and latent heat storage, or combination of these both [1]. The energy is stored in the latent heat storage systems regarding the phase change of materials as a constant or nearly constant temperature. It is critically reviewed in several recent publications that the PCM thermal storage system is one of the most efficient heat storage methods because it provides a considerable amount of energy during the charging and discharging process compared to that of the conventional sensible heat energy storage.

This means that the latent heat storage system using a PCM requires a much smaller volume of materials to store a certain amount of energy. Recently, several studies have been carried out to study the thermal behaviour of PCM thermal storage systems using FLUENT. Al-Maghalseh [2] conducted a comprehensive review of the heat transfer enhancement methods in the thermal storage systems using PCM. Several techniques of heat transfer intensification methods were discussed in details, including both experimental and numerical studies for fins, fins materials and geometry, filling materials, nano-fluid, nano-particles, microencapsulation and thermal conductivity enhancement. Another study by the same author [3] numerically studied the effects of natural convection on the heat transfer of the

PCM thermal storage system. He found that the natural convection has a considerable effect on the heat transfer inside the storage system, and therefore reducing the total melting time of the PCM. A detailed review of energy storage using PCMs has been performed in [4–8] as well as Perez-Raya [9], Joybari [10], Riahi [11], Kozak [12], Liu [13], and Mario [14].

Jian et al. [15] developed a numerical model to predict the transient thermal behaviour during charging and discharging processes of a latent thermal storage unit involving a triplex concentric tube with the PCM filling the middle channel Freeman et al. [16] carried out some investigations into a small-scale solar organic ranking cycle (ORC) with integrated PCM thermal energy storage(TES) unit. The system was examined for selected months in the contrasting climates of Cyprus and the UK. The performance indicator of the ORC engine and the required TES volume with and without the PCM are compared and discussed. It was found that the system with evacuated flat-plate collectors has a better performance compared with using low-cost evacuated-tube heat-pipe collectors. Furthermore, using PCMs for the TES shown better performance and a smaller equivalent storage volume than water.

Dal Magro et al. [17] used the PCM to improve the efficiency of the ORC system operating under thermal power fluctuations. He found that using the PCM allows the capacity factor to increase from 38–52% and the average thermal efficiency to increase from 15.5% to 16.4%. Sagar et al. [18] developed a numerical model for ORC based solar thermal power plant integrated with latent heat thermal energy storage system. Shell and tube latent heat thermal storage system was designed to generate 200 kW during the discharging process. However, the overall performance of the solar thermal power plant was evaluated for ten days of operation. Another study by Manfrida et al. [19] developed a simulation model for PCM thermal storage system coupled with solar-powered ORC. The study examined the thermal performance of the system over several conditions. Further, a case study for the operating of the system during one week was numerically examined. The results clearly showed that the system we able to provide power in 78.5% of the time, with weekly averaged efficiencies of 13.4% for the ORC unit, and 3.9% for the whole plant. A simple numerical method, called, the temperature and thermal resistance iterations, was used in the numerical calculation. The data from the numerical model was then compared with experimental results, and a good agreement was observed. Ho and Chen [20] also developed a numerical model for the melting of ice around a horizontal isothermal cylinder. The model's results were compared with experimental data published by White in [21], and a good agreement was found. It was concluded that the melting process of ice is strongly affected by the changing recirculation occurring in the molten water. Another numerical model of melting around a horizontal pipe was developed by Rieger et al. [22]. The numerical solution was obtained for Rayleigh numbers (Ra) up to 1.5×10^5, Stefan numbers in the range of $0.005 \leq Ste \leq 0.08$, and for Pr=50. It was found that the natural convection is the dominant process in the heat transfer mechanism throughout almost the entire melting process.

Trp studied the transient heat transfer in the shell-and-tube thermal storage system in an experimental and numerical study [23, 24]. He developed a mathematical model based on the non-isothermal phase transition, and it was implemented as a FORTRAN computer code. The numerical results were validated with experimented data, and it was concluded that heat transfer from the HTF to the PCM was low due to the large Prandtl numbers of the HTF. Therefore, a large amount of heat was carried downstream with the HTF, whilst a small amount of heat was transferred to the PCM upstream. The same author [25] numerically investigated the effects of several geometrical parameters and different HTF operational conditions on heat transfer during both melting and solidification processes by measuring the transient temperature distribution of the HTF, PCM and tube wall.

This chapter presents the results of the 3-D CFD modelling of the PCM in the shell-and-tube thermal storage system. Then, the numerical results obtained by the CFD were compared with the experimental and numerical data from the literature. Finally, a detailed parametric study of heat transfer processes in the melting PCM was carried out and results were discussed.

2. Validation of the FLUENT model with the experimental case study by Lacroix

Lacroix [26], conducted a series of experiments to study the heat transfer performance of the shell-and-tube thermal storage unit using PCM. The PCM installed on the shell side, while the Heat Transfer Fluid (HTF) flowing inside the tube. The effects of several thermal and geometric parameters on the heat process were investigated. The schematic diagram of the model is presented in **Figure 1**. The PCM fills the shell with the diameter of D_e, whereas the HTF flows through the tube with a diameter of Di. The PCM is commercially available material *n-Octadecane*. The thermophysical properties of the *n-Octadecane* are presented in **Table 1**.

Figure 2 illustrates the test unit scheme. Two concentric tubes were used. The inner tube (D_i = 12.7 mm, D_o = 15.8 mm, and L = 1 m) is made of copper, and

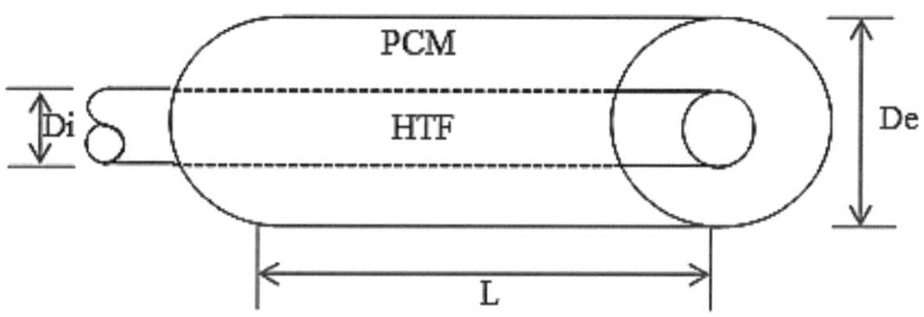

Figure 1.
Schematic representation of the test unit.

Properties	Value
Liquid density, ρ_l	814 kg/m^3
Solid density, ρ_s	814 kg/m^3
Liquid Thermal conductivity, k_l	0.148 W/(m°C)
Solid Thermal conductivity, k_s	0.358 W/(m°C)
Liquid specific heat, Cp_l	2200 J/Kg °C
Solid specific heat, Cp_s	1900 J/Kg °C
Latent heat, L	243.5 KJ/Kg
Viscosity, μ	3.878×10^{-3}Pa s
Thermal expansion coefficient, β	0.00091 1/K
Melting Temperature, T_m	300.7 K

Table 1.
Thermo-physical properties of the n-Octadecane *[27–29].*

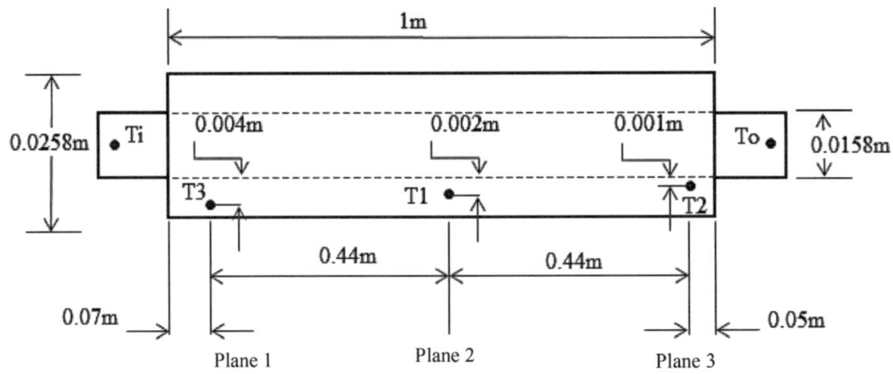

Figure 2.
Experimental test unit.

outside tube (D_i = 25.8 mm, and L = 1 m) is made of Plexiglas. Thick pipe insulation (Rubates Armstrong Armflex II) was used to isolate the system. The space between two tubes was filled with *N-Octadecane as a PCM,* while water was used as an HTF. An electrical heater inside a tank was used to maintain the inlet temperature of the HTF. Then, the HTF was circulating inside the copper tube with the mass flow ranging from 0.03 to 0.07 kg/s. Three thermocouples were used to record the temperature inside the PCM at different locations. Further, two thermocouples being used to record the inlet and outlet temperature of the HTF. A data acquisition unit was used to record the thermocouples signals into a PC. Finally, the storage unit was positioned vertically to depress the natural convection effects on the heat transfer inside the system.

For the validation purpose, Lacroix's experiments were numerically restudied using the ANSYS FLUENT software. In the preliminary simulations, different grid sizes and time steps were carefully examined to obtain computational grid convergence. The computational grid was constructed using 282504 hexahedral elements and boundary layers were used surrounding the pipe. Transient simulations were run using the *k-epsilon* turbulence model and the time step used in calculations was set to 0.1 s. To study the phase change phenomena in the PCM, the solidification/melting model was enabled. The first-order upwind spatial discretization and the pressure solver with the PRESTO algorithm for pressure–velocity coupling were selected to obtain a converged solution. Convergence criteria were established by setting the absolute residual values to 10^{-6} for energy and 10^{-3} for all other variables. Zero heat flux boundary conditions were set on all sides of the shell. The mass flow rate and temperature of the HTF were specified at the inlet of the copper pipe. The mathematical formulations for solving PCM related problems have been categorized [30] as fixed grid, variable grid, front-fixing, adaptive grid generation, and enthalpy methods. Two methods are used to analyse the heat transfer in solid-liquid PCMs. These are the temperature-based and enthalpy-based methods. In the former, temperature is considered to be a single dependent variable. The energy equations for both solid and liquid are formulated separately; and thus the solid-liquid interface positions can be tracked easily to achieve an accurate solution for the problem [31, 32].

An enthalpy-porosity method is used for modelling the solidification/melting process [33]. This technique is described in detail by Voller and Prakash [32, 34].

The energy conservation equation for this case is written as:

$$\frac{\partial}{\partial t}(\rho H) + \nabla.\left(\rho\vec{v}H\right) = \nabla.(k\nabla T) + S \tag{1}$$

The enthalpy of the material is calculated as the sum of the sensible heat, h, and latent heat, ΔH:

$$H = h + \Delta H \tag{2}$$

The sensible heat is calculated as:

$$h = h_{ref} + \int_{T_{ref}}^{T} c_p dT \tag{3}$$

The latent heat is also calculated as:

$$\Delta H = \beta_l L \tag{4}$$

The liquid fraction, β_l, can be calculated as:

$$\begin{aligned} &\beta_l = 0, when\ T < T_{solid} \\ &\beta_l = 1, when\ T > T_{solid} \\ &\beta_l = \frac{T - T_{solid}}{T_{liquid} - T_{solid}}\ \ if\ T_{solid} < T < T_{liquid} \end{aligned} \tag{5}$$

The solid and liquid temperatures are also calculated as

$$T_{solid} = T_{melt} + \sum_{solutes} K_i m_i Y_i \tag{6}$$

$$T_{liquid} = T_{melt} + \sum_{solutes} m_i Y_i \tag{7}$$

The source term in the momentum equation can be written as [33]:

$$S = \frac{(1\ \ \beta)}{(\beta_l{}^3 + \epsilon)} A_{mush} \left(\vec{v} - \vec{v}_p\right) \tag{8}$$

Due to Darcy's law damping terms as a source term are added to the momentum equation because of the effect of phase change on convection, whereas ε is a small constant number (0.001) used to prevent division by zero and A_{mush} is the mushy zone constant. Values between 10^4 and 10^7 are recommended for most computations [33]. In the present study, the mushy zone was set to 10^5. $\vec{v}_p$ is the solid velocity due to pulling solidification materials out of the domain; and in the present study, pull velocities are not included in the solution and so $\vec{v}_p$ is set to zero. More details about the numerical model can be found at [32].

The liquid velocity can be calculated by the following Equation [33]:

$$\vec{v}_{liq} = \frac{\left(\vec{v} - \vec{v}_p(1 - \beta_l)\right)}{\beta_l} \tag{9}$$

The validation of the CFD model was carried out by comparing numerical results from ANSYS FLUENT to experimental data obtained by Lacroix [26]. The comparison was carried out for three different cases during the melting process. These are for three different HTF inlet temperatures above the melting temperature of *n-Octadecane* by 5,

10 and 20 K. The HTF mass flow rate was maintained at a constant value of 0.0315 Kg/s. **Figures 3-5** show the temporal temperature variations in the experiment and CFD data at locations T1 (h = 0.51 m, r = 0.002 m) and T2 (h = 0.95 m, r = 0.001 m) inside the PCM. It is clearly shown that the predicted numerical results of the temperature follow the experimental trend in Lacroix [26]. The main discrepancies between the

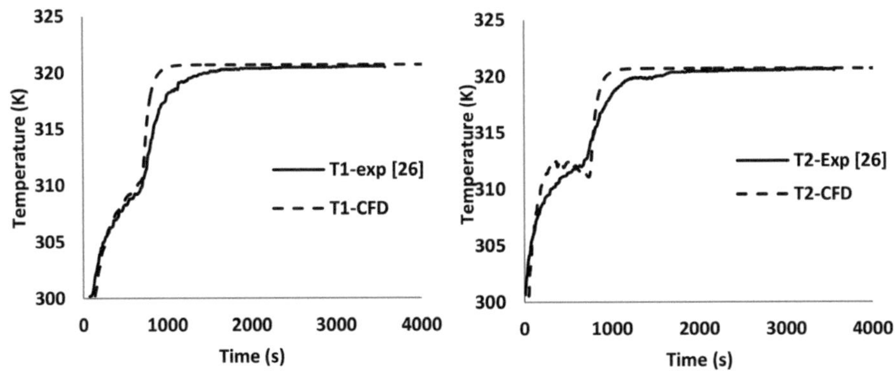

Figure 3.
The variation of the predicted and experimental temperature at locations T1 and T2 ($T_{in} = T_m + 20$ K), the mass flow rate = 0.0315 kg/s.

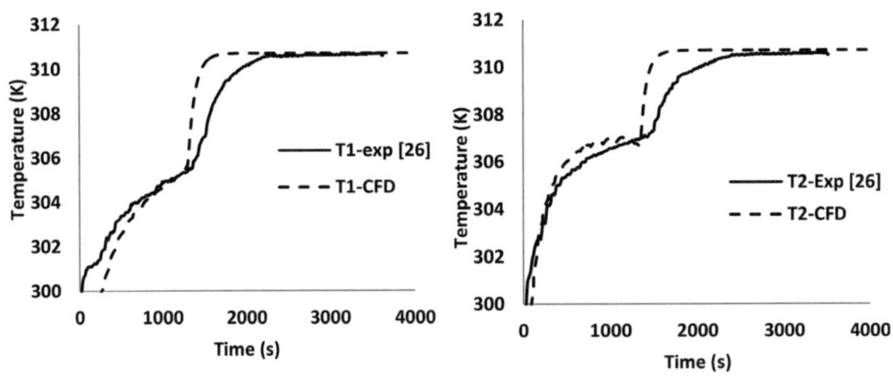

Figure 4.
The variation of the predicted and experimental temperature at locations T1 and T2 ($T_{in} = T_m + 10$ K), the mass flow rate = 0.0315 kg/s.

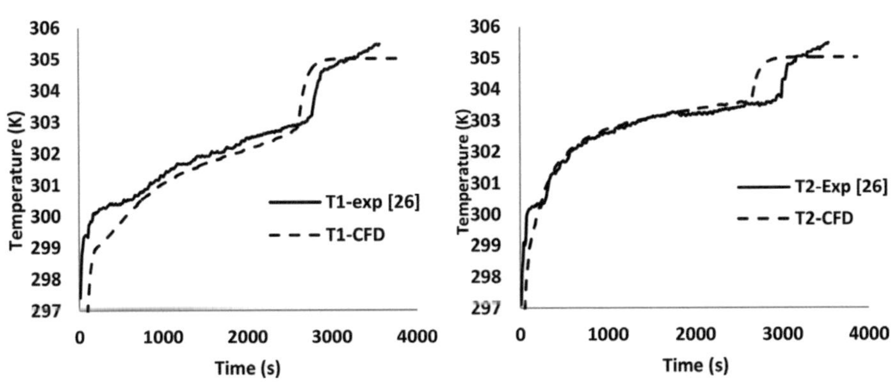

Figure 5.
The variation of the predicted and experimental temperature at locations T1 and T2 ($T_{in} = T_m + 5$ K), the mass flow rate = 0.0315 kg/s.

numerical and experimental results can be attributed to the measurements uncertainties and the difference in the PCM physical properties in the solid and liquid phase. Furthermore, CFD numerical results on the evolution of the liquid fraction in the PCM were compared with calculations of Lacroix [26]. **Figure 6** shows the variation of molten volume fraction of the PCM in the test unit as a function of time. The outer diameter (D_e) of the storage unit is 22 mm, the inside tube diameter (Di) is 12.7 mm, and the storage length (L) is 1 m. The HTF mass flow rate in simulations ranges from 1.5×10^{-4} to 1.5×10^{-2} and the HTF inlet temperature was 20 K above the PCM melting temperature. It can be seen in **Figures 7-9**, the current CFD results

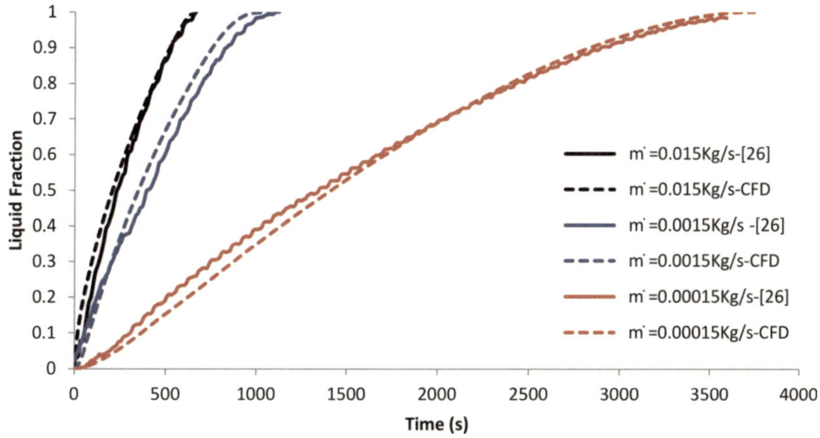

Figure 6.
The CFD liquid fraction variation in time against the numerical liquid fraction from [26] at several mass flow rate.

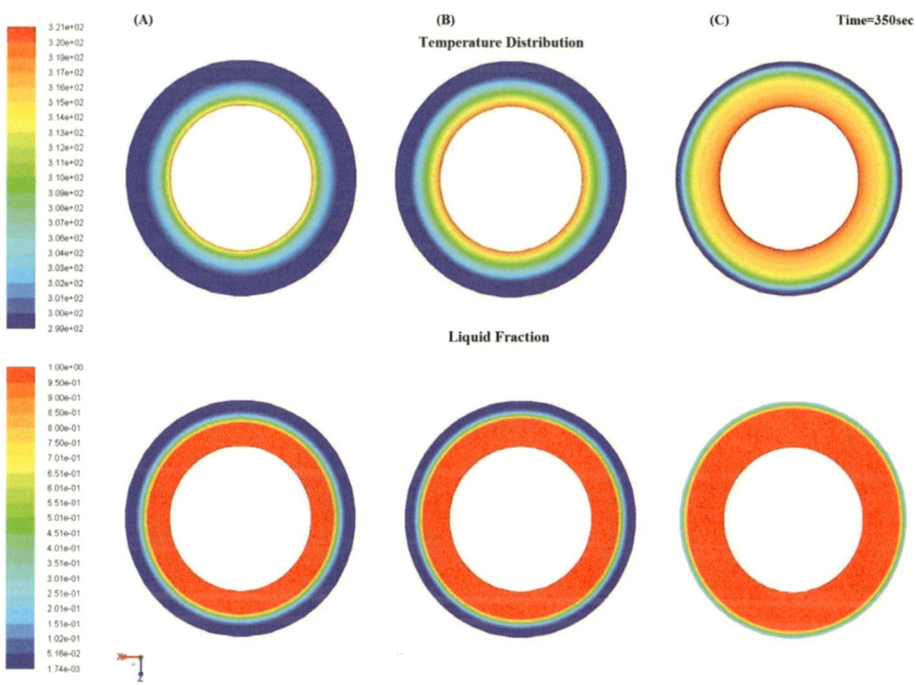

Figure 7.
Temperature distribution in the PCM for (A) bottom section plane, (B) middle section plane, and (C) top section plane (T_{in} = 320 K, HTF mass flow rate = 0.0315Kg/s., the elapsed time is 350 sec).

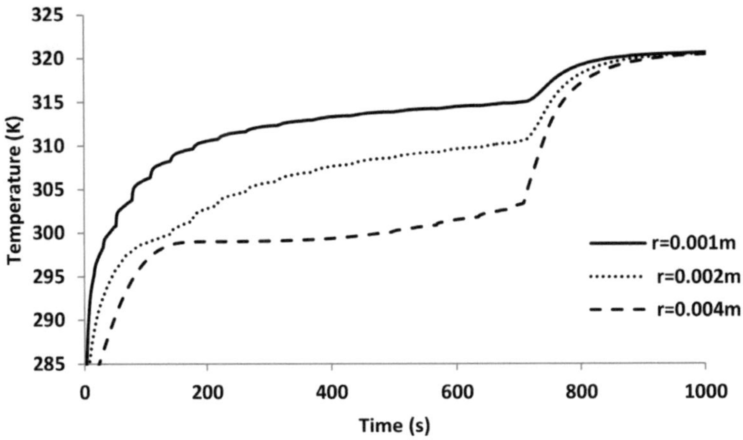

Figure 8.
PCM temperature versus time at the different radial positions: R = 0.001, 0.002, and 0.004 m from the axis of the computational domain at y = 0.5 m, T_{in} *= 320 K, HTF mass flow rate = 0.0315Kg/s.*

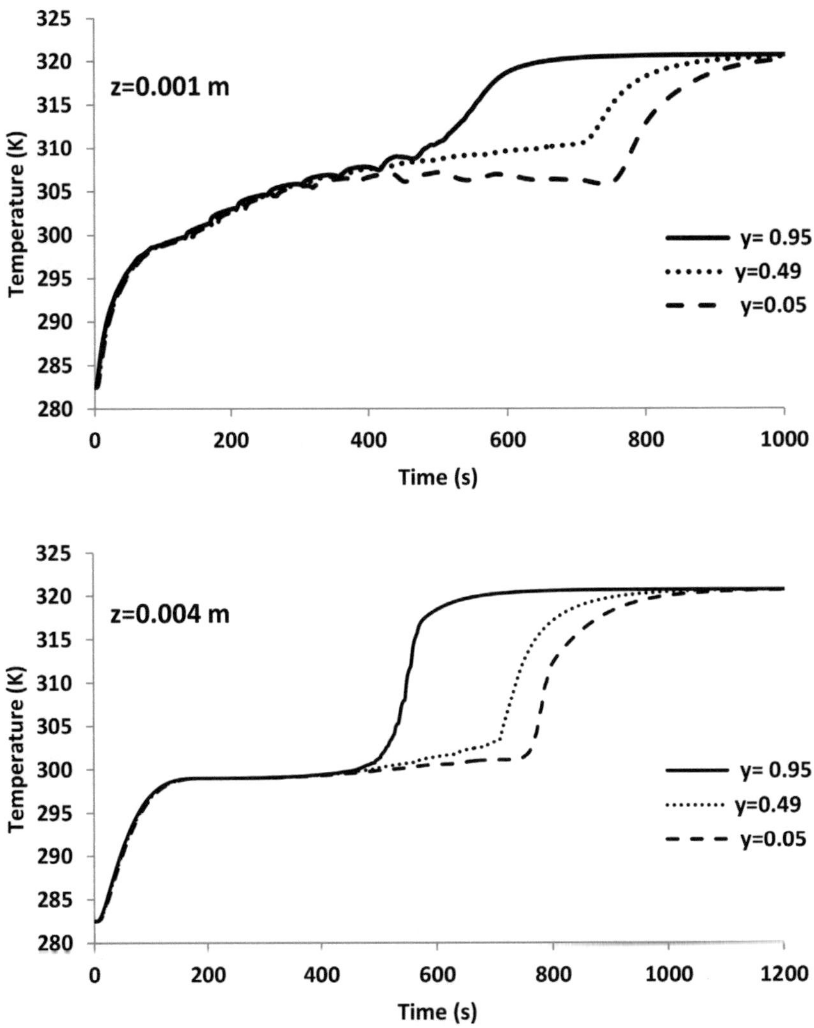

Figure 9.
PCM temperature versus time along the axis (y = 0.05, 0.49, and 0.95 m) for the radial distances z = 0.001 and 0.004 m, T_{in} *= 320 K, HTF mass flow rate = 0.0315Kg/s.*

are in a very good agreement with calculations of Lacroix [26]. In general, comparison of CFD results with results presented in Lacroix [26] demonstrated that the developed CFD model accurately describes processes taking place in the experimental test rig and therefore can be used with confidence for further transient heat transfer simulations in the shell-and-tube latent thermal storage unit.

3. Heat transfer performance

The heat transfer performance of the storage unit was numerically examined during the charging process. **Figure 7** shows the numerical results of the temperature and liquid fractions along the storage unit axis during the charging process when the elapsed time is 350 seconds (the inlet temperature of the HTF is 320 K, the mass flow rate is 0.0315 Kg/s).

It can be seen that the highest temperature of the PCM can be observed at the domain's top region close to the inlet of the HTF and can rise gradually in the regions closed to at the vicinity of the tube walls. Therefore, the top part of the domain converted into a liquid first and later on, melting expands to lower regions on the domain. **Figure 8**, illustrates the temperature variation in the PCM for three different radial locations. As expected, the higher temperatures are noticed in the regions close to the surface of the wall with the HTF, where the melting process takes place first. The temperature variations along the axis at the outer surface of the HTF tube are shown in **Figure 9**. It can be seen that higher temperatures exist at the domain's top. This is mainly because the HTF flows from the top to the bottom and thus the temperature rises faster at the vicinity of the tube walls close to the inlet. At this period, the effect of natural convection is not profound yet.

The effect of HTF inlet temperature on the melting process is demonstrated in **Figure 10**. It can be seen that the inlet temperature of the HTF considerably affects the rate of the melting and the PCM temperature distribution. The increase in the inlet temperature of the HTF leads to a rise in the temperature difference between the tube walls and the bulk of the PCM and thus enhance the heat transfer rate. It results in the faster rise of the liquid fraction and decreases the total melting time. **Figure 11** shows that the time for completion of melting for the HTF inlet temperature of 305 K is 2974 s, for 310 K this time is reduced to 1617s and finally, for 320 K, the time of melting is only 963 s. Therefore, the total melting time is reduced

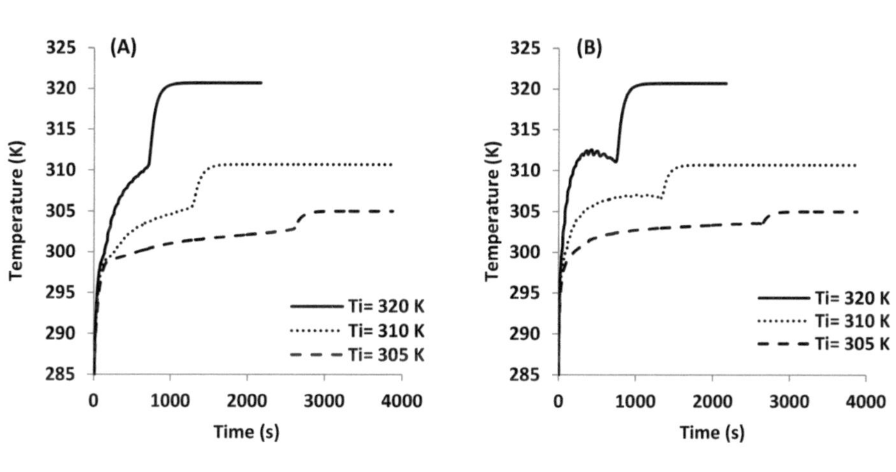

Figure 10.
Effect of the HTF inlet temperature on the melting process: (A) location T1 (h = 0.51 m, r = 0.002 m) and (B) location T2 (h = 0.95 m, r = 0.001 m), T_{intial} *= 282.5 K, HTF mass flow rate = 0.0315Kg/s.*

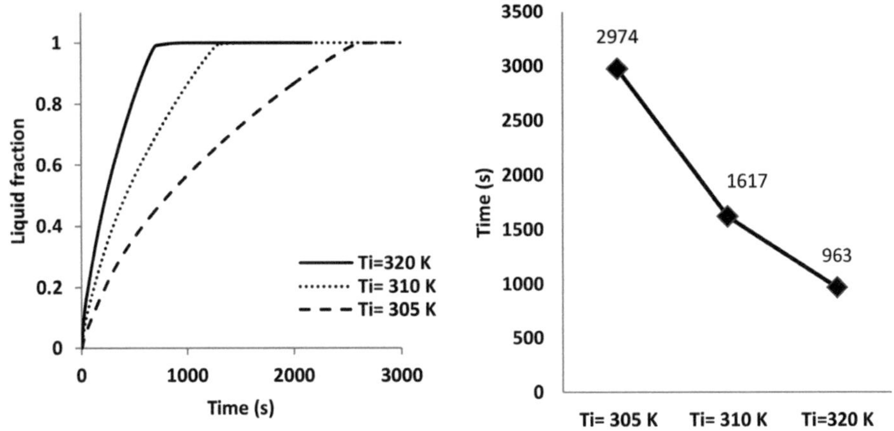

Figure 11.
The effect of the inlet temperature of the HTF on the formation of the liquid fraction of the PCM and on the melting time.

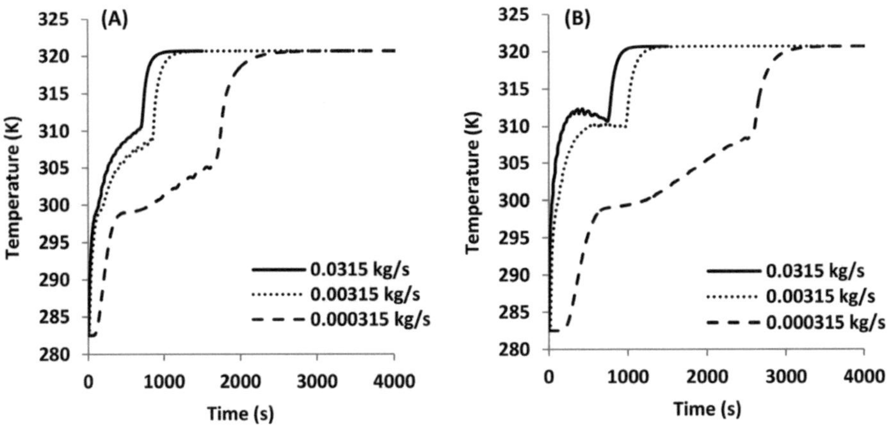

Figure 12.
Effect of the HTF flow rate on the melting process. The PCM temperature at locations (A) T1 (h = 0.51 m, r = 0.002 m) and (B) T2 (h = 0.95 m, r = 0.001 m), T_{intial} = 282.5 K, HTF mass flow rate = 0.0315Kg/s.

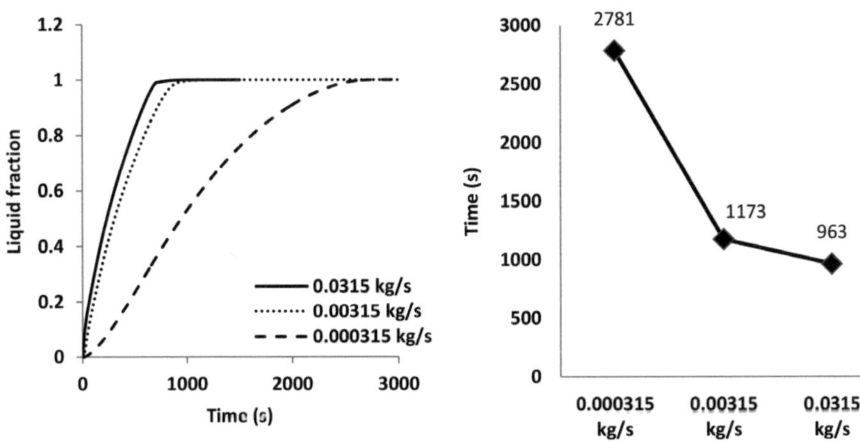

Figure 13.
The effect of the HTF flow rate on the liquid fraction formation and the melting time, T_{intial} = 282.5 K, T_{inlet} = 320 K.

Figure 14.
Temperature distribution and melting process at the bottom of the computational domain for the different HTF mass flow rates and inlet temperatures (elapsed time is 850 s).

approximately by 68% when the inlet temperature is increased from 305 to 320 K and by 45.6% when the inlet temperature is increased from 305 to 310 K.

Figure 12 demonstrates the effect of the HTF flow rate on the meting process. It can be seen that the flow rate accelerates the melting process due to the increased heat transfer rate. It can be seen from **Figure 13** that when the flow rate increases from 0.000315 to 0.00315 kg/s, the PCM melting time is reduced from 2781 to 1173 s (reduction by 57%). Also, the melting time is reduced by 17.8% when the mass flow rate increases from 0.00315 to 0.0315 kg/s. The effect of the HTF mass flow rate rise is less profound when compared with the effect of the rise in the HTF inlet temperature. This is demonstrated in **Figure 14** using temperature distribution counters and the PCM fluid fraction evolution diagrams.

4. Effect of natural convection

To study the effect of the natural convection, the system needs to be installed in the horizontal position. Several points were created inside the computational domain to monitor the variation of the temperature inside the PCM during numerical CFD modelling. These monitoring points are placed in three planes, which are perpendicular to the axis of the system and located at distances of 0.07, 0.51 and 0.95 m from the front of the system (see **Figure 1**). **Figure 15** indicates the locations of monitoring points in the plane at a distance of 0.07 m from the front of the system.

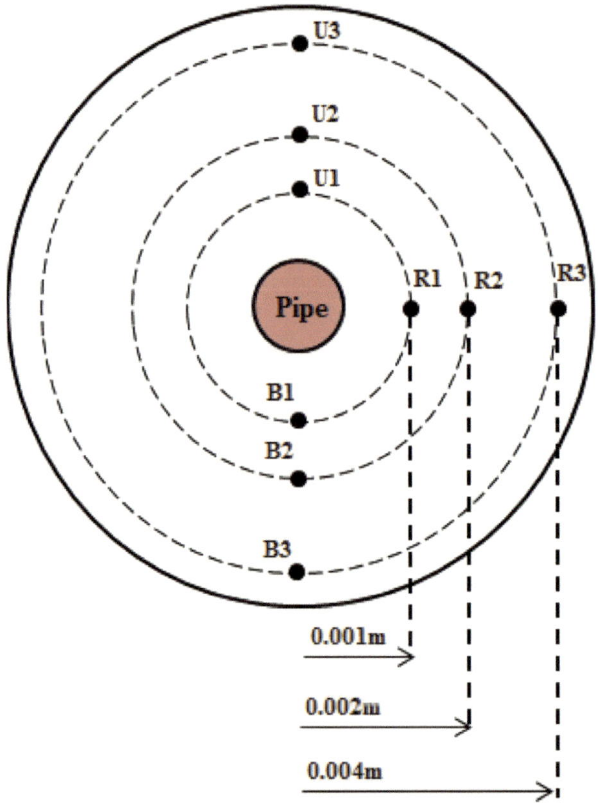

Figure 15.
Locations of the monitoring points around the pipe.

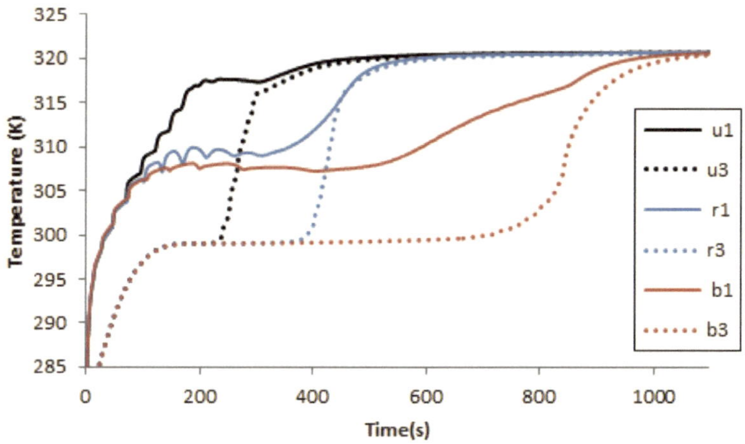

Figure 16.
Variation of the base PCM temperature with time.

Figure 16 shows the temperature variation at some of the monitoring points at the front plane of the system with the PCM for the case when the inlet temperature of the HTF is 320.7 K and the mass flow rate is 0.0315 kg/s. It can be seen that initially, the temperature increases rapidly from 280 to 299 K due to the heat transfer from the pipe walls to the solid PCM by conduction. The temperatures at monitoring points u1, r1, and b1 rise more rapidly due to their proximity to the pipe. Temperatures in monitoring points u3, r3, and b3 rise considerable slower since these points are located on the edge of the storage unit. During the melting process the temperature increases from 299 to 300.7 K and equalise at all monitoring points. Initially, during the melting process, a thin liquid layer is formed between the pipe and the solid PCM. Gradually, the solid–liquid interface expands in the axial and radial directions and the melting then is dominated by natural convection in the PCM's liquid regions. The melting process is intensified in the upper regions of the container, resulting in higher temperature recordings at the top of the computational domain (point u1).

The velocity vectors in the liquid pure PCM are shown in **Figure 17** for the elapsed time of 300, 400 and 550 s. It can be seen that the molten PCM ascends upwards from the top regions at the centre of the unit and then after cooling flows downwards to complete the natural convection circle. The convection is intensified as the liquid fraction volume increases. The velocity magnitude gradually decreases in time due to a reduction in the temperature difference in the molten PCM. These results are in good agreement with the results of several experimental and numerical investigations [35–37].

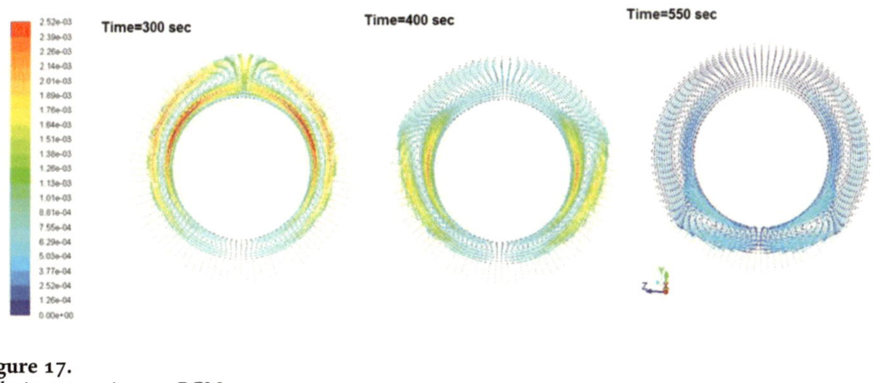

Figure 17.
Velocity vectors in pure PCM.

5. Average heat transfer coefficient

The local heat transfer coefficient could not be estimated accurately for the present thermal storage system as there is a temperature difference between the outer surface of the HTF pipe and the PCM along with both axial and radial directions. Consequently, the average heat transfer coefficient for the melting process is calculated instead using the following Equations [38] for the temporal heat transfer coefficient

$$h_p = \frac{q}{A \times \Delta T(LMTD)} \tag{10}$$

where the surface area of HTFP is calculated using the following equation:

$$A = \pi D_o l \tag{11}$$

The heat transfer rate (q) in the thermal storage can be calculated through the HTF's enthalpy reduction rate in the HTF [39]. This enthalpy reduction can be calculated through the following equation:

$$q = \dot{m} c_p (T_i - T_o) \tag{12}$$

where, $\dot{m}$ is the HTF mass flow, T_i and T_o are the HTF's inlet and outlet temperatures respectively.

The average heat transfer coefficient is:

$$\overline{h_p} = \frac{Q_{total}}{A \times \Delta T(LMTD) t_n} \tag{13}$$

$$\Delta T(LMTD) = \frac{(T_{w1} - T_{u1}) - (T_{w2} - T_{u2})}{[\ln (T_{w1} - T_{u1})/(T_{w2} - T_{u2})]} \tag{14}$$

where T_{u1} and T_{u2} are the PCM temperature at points u1 (first measurement plane) and u2 (last measurement plane); T_{w1} and T_{w2} are the pipe wall temperatures at the same corresponding measurement planes.

Finally, the temporal Nusselt number (Nu) is used to quantify heat transfer and this can be calculated using the following equation:

$$Nu = \frac{h_p r_{eq}}{k_{pcm}} \tag{15}$$

The time-averaged Nusselt number $\overline{Nu}$ is defined as

$$\overline{Nu} = \frac{\overline{h_p} r_{eq}}{k_{pcm}} \tag{16}$$

Different cases were analysed with various system's geometrical and thermophysical parameters. To generalise results it is vital to characterise them in the dimensionless form. More details about the dimensional parameters calculation and analysis can be found at [32].

The heat transfer coefficient values were calculated for the top, side and bottom regions of the storage unit for several cases during the melting processes.. **Figure 18** present results on the heat transfer coefficient changing as a function of time for the PCM during the case when the inlet temperature of the HTF is 320.7 K and the mass

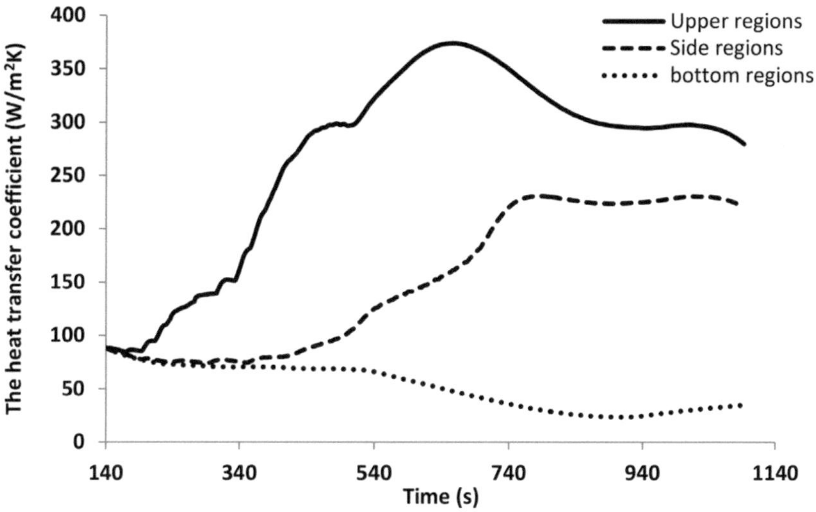

Figure 18.
The heat transfer coefficient variation at the top, side, and bottom regions of the system.

flow rate is 0.0315 kg/s. It can be seen that the heat transfer coefficients for both top and side regions increase with time. This agrees well with the initial solid PCM's temperature rise followed by the melting process. The higher heat transfer coefficient values can be observed in the top regions of the system. This can be attributed to the effect of natural convection. For the longer elapsed times, the heat transfer value stabilises. This is because the temperature distribution becomes more established within the PCM body when the system reaches the steady-state operation. At this stage, the majority of the PCM is melted and a very small part of it, which is close to the bottom regions of the unit, maybe in the solid-state.

The average heat transfer coefficient values for various HTF inlet temperature and flow rates were obtained and used to calculate the Nusselt number for the pure PCM. The flow rates of the HTF considered were 0.000315, 0.00315, 0.005, and 0.0063 Kg/s. The HTF inlet temperatures used in the modelling are 305, 310, and 320 K. To derive generic heat transfer correlations, the Nusselt numbers were calculated at the top, side and bottom regions of the storage unit. Thereafter, the average Nusselt number value for all regions is calculated. **Figures 19-21** show the results of these CFD modelling. As it can be seen in **Figure 19**, the Nusselt number increases with the rise in the Stefan number (see Eq. 18) which is proportional to the difference between the inlet temperature of the HTF and the melting temperature of the PCM (the increase in the inlet temperature of HTF increases the Stefan number). Similar observations can be made concerning the Rayleigh number (see Eq. 19), see **Figure 20**. The Rayleigh number is also proportional to the difference between the inlet temperature of the HTF and the melting temperature of the PCM.

$$Ste = \frac{Cp_{pcm}\Delta T_H}{\lambda_{pcm}} \quad (17)$$

$$Ra = \frac{g\beta Cp_{pcm}\rho_{pcm}\Delta T_H r_{eq}^3}{\mu_{pcm}k_{pcm}} \quad (18)$$

$$Fo = \frac{(\alpha_{pcm}t_m)}{(r_{eq})^2} \quad (19)$$

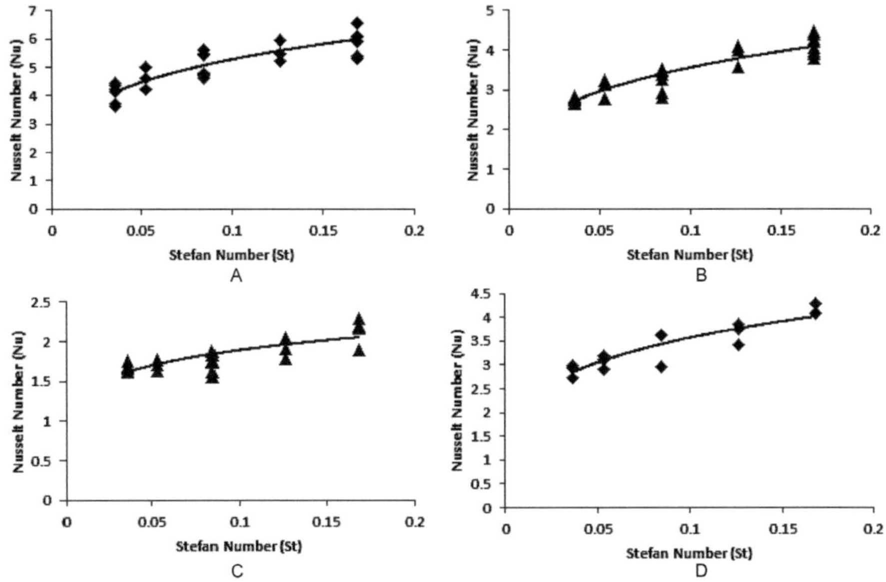

Figure 19.
Variation in the Nusselt number as a function of Stefan number for (A) upper regions (B) side regions (C) bottom regions (D) the average Nusselt number (in the system).

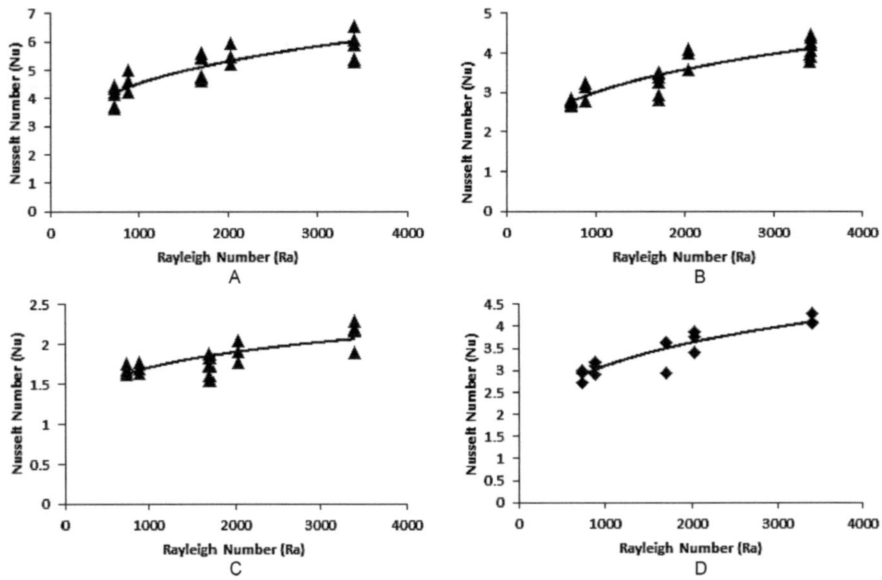

Figure 20.
Variation in the Nusselt number as a function of Rayleigh number for (A) upper regions (B) side regions (C) bottom regions (D) the average Nusselt number (in the system).

Figure 21 show that the Nusselt number decreases with the rise in the Fourier number (see Eq. 20). The Fourier number is proportional to the melting time. However, an increase in the inlet temperature and flow rate of the HTF will lead to a decrease in the total melting time, and thus, reduce the Fourier number.

The average Nusselt numbers in the system are presented in **Figure 19D, 20D,** and **21D**. This data was used to derive the Nusselt number correlations for the pure

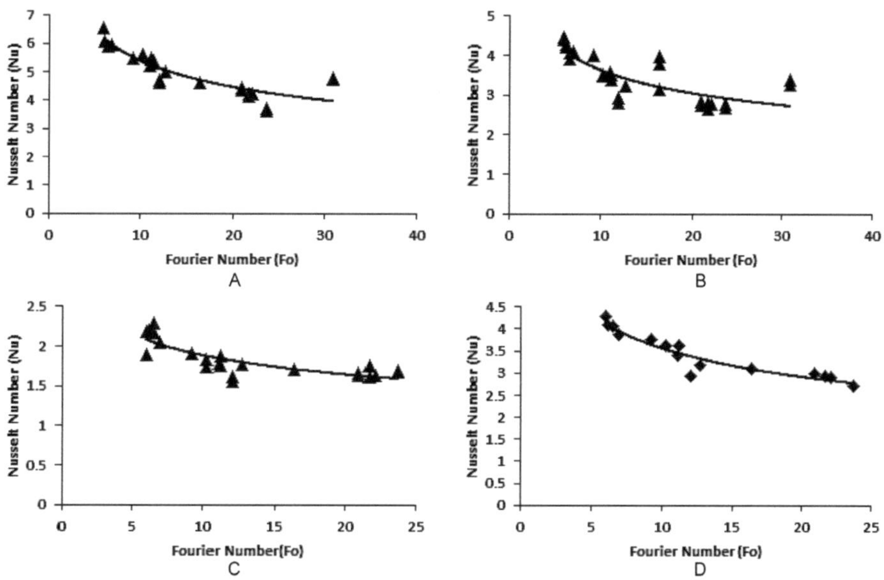

Figure 21.
Variation in the Nusselt number as a function of Fourier number for (A) upper regions (B) side regions (C) bottom regions (D) the average Nusselt number (in the system).

PCM. The correlations are derived as a function of the Stefan, Fourier, and Rayleigh numbers.

The Nusselt number for the pure PCM is ($R^2 = 0.8524$)

$$Nu = 2.9883\,(Ste)^{0.0758}(Fo)^{-0.095}(Ra)^{0.0759} \quad (20)$$

The correlation between the numerically obtained and calculated using Eqs. (20) is shown in **Figure 22**. It can be seen that the Nusselt number varies between 2 and 4.3 for the system under investigations. The lower Nusselt numbers are for the low HTF inlet temperature, and the high Nusselt numbers values are obtained at the high HTF inlet temperatures. The total melting time of the PCM can then be calculated using the following formula ($R^2 = 0.966$)

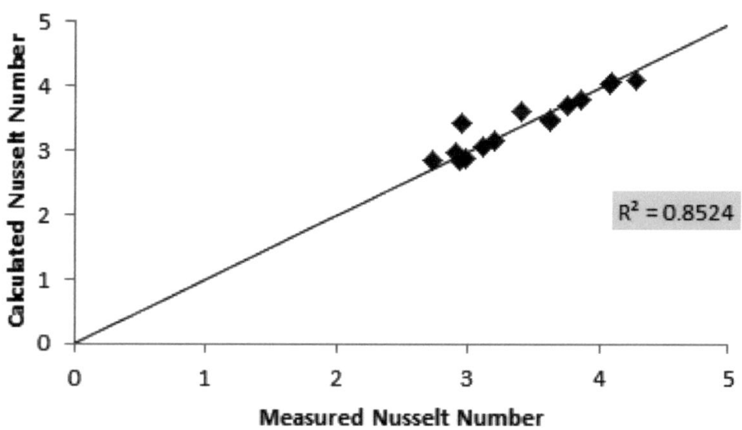

Figure 22.
The correlation between numerically obtained and calculated using Eq. (20) Nusselt numbers - the pure PCM.

$$t_m = \frac{\rho D}{1.056\bar{h}} \left[\frac{c_p(T_m - T_{\text{intial}}) + \lambda_{pcm}}{(T_{inlet} - T_m)} \right] \tag{21}$$

6. Conclusions

The 3D CFD simulation model was developed for shell-and-tube thermal storage system using FLUENT/ANSYS. For the validation purpose, the results from the numerical model were compared with the experimental results of Lacroix [26]. The results show that a considerable agreement between the numerical and experimental results. Therefore, the results demonstrate that the developed CFD model accurately describes processes taking place in the experimental test rig and therefore can be used with confidence for further transient heat transfer simulations in the shell-and-tube latent thermal storage unit. 3D CFD simulations were performed for a range of the HTF inlet temperature values and its mass flow rate. Thereby, the results were used to derive Nusselt number correlations as a function of Stefan, Rayleigh and Fourier numbers to take into account the effect of all design and operational conditions. The Nusselt number for the system with the pure PCM was found to be ($R^2 = 0.8524$)

$$Nu = 2.9883\,(Ste)^{0.0758}(Fo)^{-0.095}(Ra)^{0.0759}$$

The total melting time of the PCM can be estimated by ($R^2 = 0.966$)

$$t_m = \frac{\rho D}{1.056\bar{h}} \left[\frac{c_p(T_m - T_{\text{intial}}) + \lambda_{pcm}}{(T_{inlet} - T_m)} \right]$$

Nomenclature

A	Area (m^2)
A_{Mush}	Mushy zone constant
C_P	Specific heat (J/Kg°C)
D	Diameter
g	Gravitational acceleration
Gr	Grashof number
h	Sensible enthalpy (J)
H	Enthalpy
H	Height
h_{ref}	Reference enthalpy
k	Thermal conductivity (W/m°C)
k_{eff}	Effective thermal conductivity
L	Latent heat (Kj/Kg)
L	Length
LMTD	The logarithmic mean temperature difference
m	Mass in each space (Kg)
$\dot{m}$	Mass flow rate
Nu	Nusselt number
Pr	Prandtl number
q	Heat transfer rate
r	Radial coordinate
Ra	Rayleigh number

Re	Reynolds number
S	Source term
S_h	The heat from any volumetric sources
S_m	Mass source
Ste.	Stefan number
t	Time (s)
T	Temperature
T_{ref}	Reference temperature
u	X velocity components (m/s)
v	Y velocity components (m/s)
v_n	Normal components of the velocity of the interface
v_r	Radial velocity component
v_x	Axial velocity component
w	Z velocity components (m/s)
W	Width
x	Axial coordinate
Y_M	Contribution of the fluctuating dilatation incompressible turbulence to The overall dissipation rate

Subscripts

L	Liquid
eff	Effective
f	Fluid
i	Inlet
m	Melting
o	Outlet
s	Solid
T	Total
w	Wall

Greek symbols

ρ	Density (Kg/m^3)
β	Coefficient of thermal expansion (1/K)
μ	Viscosity (Kg/ms)
β_t	Liquid fraction
α	Permeability
ϕ	Volumetric fraction

Author details

Maher Mohammad Al-Maghalseh[1,2]

1 College of Engineering, Palestine Polytechnic University, Hebron, Palestine

2 Faculty of Engineering and Environment, Northumbria University, Newcastle, UK

*Address all correspondence to: maherm@ppu.edu

References

[1] A. Abhat, "Low temperature latent heat thermal energy storage: Heat storage materials," *Solar Energy*, vol. 30, pp. 313–332, 1983.

[2] M. Al-Maghalseh and K. Mahkamov, "Methods of heat transfer intensification in PCM thermal storage systems: Review paper," *Renewable and Sustainable Energy Reviews*, vol. 92, pp. 62–94, 2018/09/01/ 2018.

[3] M. M. Al-Maghalseh, "Investigate the Natural Convection Heat Transfer in A PCM Thermal Storage System Using ANSYS/FLUENT," *Jordan Journal of Mechanical & Industrial Engineering*, vol. 11, 2017.

[4] M. Kenisarin and K. Mahkamov, "Solar energy storage using phase change materials," *Renewable and Sustainable Energy Reviews*, vol. 11, pp. 1913–1965, 2007.

[5] M. M. Farid, A. M. Khudhair, S. A. K. Razack, and S. Al-Hallaj, "A review on phase change energy storage: materials and applications," *Energy Conversion and Management*, vol. 45, pp. 1597–1615, 2004.

[6] F. Agyenim, N. Hewitt, P. Eames, and M. Smyth, "A review of materials, heat transfer and phase change problem formulation for latent heat thermal energy storage systems (LHTESS)," *Renewable and Sustainable Energy Reviews*, vol. 14, pp. 615–628, 2010.

[7] B. Zalba, J. M. Marín, L. F. Cabeza, and H. Mehling, "Review on thermal energy storage with phase change: materials, heat transfer analysis and applications," *Applied Thermal Engineering*, vol. 23, pp. 251–283, 2003.

[8] H. Akeiber, P. Nejat, M. Z. A. Majid, M. A. Wahid, F. Jomehzadeh, I. Zeynali Famileh, *et al.*, "A review on phase change material (PCM) for sustainable passive cooling in building envelopes," *Renewable and Sustainable Energy Reviews*, vol. 60, pp. 1470–1497, 2016/07/01/ 2016.

[9] I. Perez-Raya and S. G. Kandlikar, "Numerical modeling of interfacial heat and mass transport phenomena during a phase change using ANSYS-Fluent," *Numerical Heat Transfer, Part B: Fundamentals*, vol. 70, pp. 322–339, 2016/10/02 2016.

[10] M. M. Joybari, Fariborz Haghighat, and Saeid Seddegh, "Numerical investigation of a triplex tube heat exchanger with phase change material: Simultaneous charging and discharging," *Energy and Buildings*, vol. 139, pp. 426–438, 2017.

[11] S. Riahi, W. Y. Saman, F. Bruno, and N. H. S. Tay, "Numerical modeling of inward and outward melting of high temperature PCM in a vertical cylinder," *AIP Conference Proceedings*, vol. 1734, p. 050039, 2016.

[12] Y. Kozak, M. Farid, and G. Ziskind, "Experimental and comprehensive theoretical study of cold storage packages containing PCM," *Applied Thermal Engineering*, vol. 115, pp. 899–912, 3/25/ 2017.

[13] J. Liu, C. Xu, X. Ju, B. Yang, Y. Ren, and X. Du, "Numerical investigation on the heat transfer enhancement of a latent heat thermal energy storage system with bundled tube structures," *Applied Thermal Engineering*, vol. 112, pp. 820–831, 2/5/ 2017.

[14] M. Cascetta, G. Cau, P. Puddu, and F. Serra, "A comparison between CFD simulation and experimental investigation of a packed-bed thermal energy storage system," *Applied Thermal Engineering*, vol. 98, pp. 1263–1272, 2016/04/05/ 2016.

[15] L. Jian-you, "Numerical and experimental investigation for heat transfer in triplex concentric tube with phase change material for thermal energy storage," *Solar Energy,* vol. 82, pp. 977–985, 2008.

[16] J. Freeman, I. Guarracino, S. A. Kalogirou, and C. N. Markides, "A small-scale solar organic Rankine cycle combined heat and power system with integrated thermal energy storage," *Applied Thermal Engineering,* vol. 127, pp. 1543–1554, 2017/12/25/ 2017.

[17] F. Dal Magro, M. Jimenez-Arreola, and A. Romagnoli, "Improving energy recovery efficiency by retrofitting a PCM-based technology to an ORC system operating under thermal power fluctuations," *Applied Energy,* vol. 208, pp. 972–985, 2017/12/15/ 2017.

[18] S. Lakhani, A. Raul, and S. K. Saha, "Dynamic modelling of ORC-based solar thermal power plant integrated with multitube shell and tube latent heat thermal storage system," *Applied Thermal Engineering,* vol. 123, pp. 458–470, 2017/08/01/ 2017.

[19] G. Manfrida, R. Secchi, and K. Stańczyk, "Modelling and simulation of phase change material latent heat storages applied to a solar-powered Organic Rankine Cycle," *Applied Energy,* vol. 179, pp. 378–388, 2016/10/01/ 2016.

[20] C. J. Ho and S. Chen, "Numerical simulation of melting of ice around a horizontal cylinder," *International Journal of Heat and Mass Transfer,* vol. 29, pp. 1359–1369, 1986.

[21] D. A. White, "Melting of ice and freezing of water around a horizontal cylinder," M.S Purdue University, Indiana, 1984.

[22] H. Rieger, U. Projahn, and H. Beer, "Analysis of the heat transport mechanisms during melting around a horizontal circular cylinder," *International Journal of Heat and Mass Transfer,* vol. 25, pp. 137–147, 1982.

[23] A. Trp, "A numerical and experimental study of transient heat transfer in a shell-and-tube latent heat storage unit with paraffin as a phase change material," presented at the Energy and the Environment, 2002.

[24] A. Trp, "An experimental and numerical investigation of heat transfer during technical grade paraffin melting and solidification in a shell-and-tube latent thermal energy storage unit," *Solar Energy,* vol. 79, pp. 648–660, 2005.

[25] A. Trp, K. Lenic, and B. Frankovic, "Analysis of the influence of operating conditions and geometric parameters on heat transfer in water-paraffin shell-and-tube latent thermal energy storage unit," *Applied Thermal Engineering,* vol. 26, pp. 1830–1839, 2006.

[26] M. Lacroix, "Numerical simulation of a shell-and-tube latent heat thermal energy storage unit," *Solar Energy,* vol. 50, pp. 357–367, 1993.

[27] M. Pinelli and S. Piva, "Solid/Liquid Phase Change in Presence of Natural Convection: A Thermal Energy Storage Case Study," *Journal of Energy Resources Technology,* vol. 125, pp. 190–198, 2003.

[28] J. D. Chung, J. S. Lee, and H. Yoo, "Thermal instability during the melting process in an isothermally heated horizontal cylinder," *International Journal of Heat and Mass Transfer,* vol. 40, pp. 3899–3907, 1997.

[29] V. R. Voller, "FAST IMPLICIT FINITE-DIFFERENCE METHOD FOR THE ANALYSIS OF PHASE CHANGE PROBLEMS," *Numerical Heat Transfer, Part B: Fundamentals,* vol. 17, pp. 155–169, 1990/01/01 1990.

[30] N. Özişik, *Finite Difference Methods in Heat Transfer,*: Taylor & Francis, 1994.

[31] A. A. Al-abidi, S. Bin Mat, K. Sopian, M. Y. Sulaiman, and A. T. Mohammed, "CFD applications for latent heat thermal energy storage: a review," *Renewable and Sustainable Energy Reviews,* vol. 20, pp. 353–363, 2013.

[32] M. Al-Maghalseh, " Compact solar thermal energy storage systems using phase change materials," Doctoral thesis, , Northumbria University., 2014.

[33] *ANSYS FLUENT 14.0 Theory Guide,* : ANSYS, Inc, 2011.

[34] V. R. Voller and C. Prakash, "A fixed grid numerical modelling methodology for convection-diffusion mushy region phase-change problems," *International Journal of Heat and Mass Transfer,* vol. 30, pp. 1709–1719, 1987.

[35] S. Wu, H. Wang, S. Xiao, and D. Zhu, "Numerical Simulation on Thermal Energy Storage Behavior of Cu/paraffin nanofluids PCMs," *Procedia Engineering,* vol. 31, pp. 240–244, 2012.

[36] S. Wu, D. Zhu, X. Zhang, and J. Huang, "Preparation and Melting/Freezing Characteristics of Cu/Paraffin Nanofluid as Phase-Change Material (PCM)," *Energy & Fuels,* vol. 24, pp. 1894–1898, 2010/03/18 2010.

[37] Y.-D. Liu, Y.-G. Zhou, M.-W. Tong, and X.-S. Zhou, "Experimental study of thermal conductivity and phase change performance of nanofluids PCMs," *Microfluidics and Nanofluidics,* vol. 7, pp. 579–584, 2009/10/01 2009.

[38] J. P. Holman, *Heat Transfer*: McGraw Hill Higher Education, 2010.

[39] A. Shukla, D. Buddhi, and R. L. Sawhney, "Solar water heaters with phase change material thermal energy storage medium: A review," *Renewable and Sustainable Energy Reviews,* vol. 13, pp. 2119–2125, 2009.

Chapter 5

Solidification and Melting of Phase Change Material in Cold Thermal Storage Systems

Hani Hussain Sait

Abstract

Cold thermal storage can be used to manage peak load when the energy demand is exceeding the capacities of the electric companies. Latent heat thermal storage is more effective because it requires less spacing and has higher thermal capacity than other types. Solidification and melting are taking place in CTS and need more investigation for better performance. Phase change materials properties vary and need more investigation to select the most suitable for a certain application. The analytical equations are needed for design of CTS and get the maximum efficiency out of it. Cost effectiveness is also described.

Keywords: solidification, melting, cold thermal storage

1. Introduction

The electric energy consumption in a country like Saudi Arabia reaches its peak during summer time. Most of this consumption goes to air conditioning, i.e. 75% to 90% of total electric energy production by the Saudi Electrical Companies during summer season are consumed for cooling. This put an excessive load on the electricity suppliers during summer time, at which a peak load exists as shown in **Figure 1**. The peak load or Peak demand are terms used in energy demand management describing a period in which electrical power is expected to be provided for a sustained period at a significantly higher than average supply level.

Peak load occurs usually during the day in hot countries. Most of the energy was consumed by the air conditioning for residential and commercial buildings. Load management initiatives are usually investigated by the electric companies to smooth the system load curve (Load leveling of electricity) (**Figure 1**). Many methods are suggested to handle this problem, such as utilizing renewable energy, unfixed Tarriff, use of the electrical link, and finally utilizing of energy storage systems.

Cold thermal storages were built successfully in several projects in Saudi Arabia such as, Al Mamlaka Tower in Riyadh, King Khalid Training Center in Riyadh and King Abdulaziz University Campus in Jeddah.

1.1 Concept of thermal energy storage

Unique solution to manage the peak load, that can save energy and will not cost so much is the storing of energy. The different forms of energy that can be stored include mechanical, electrical and thermal energy. Note that the Energy storage is

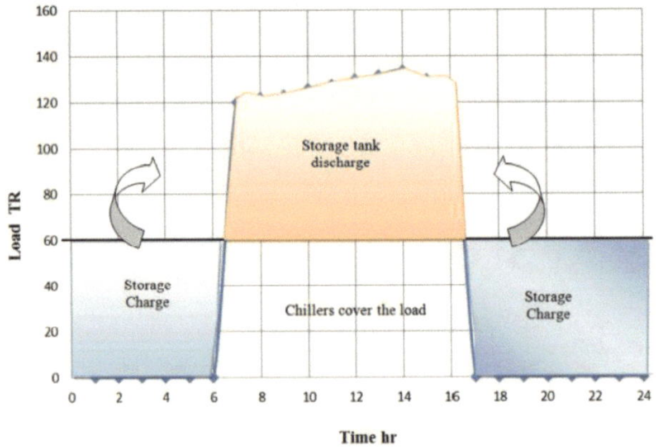

Figure 1.
Load distribution during the day (load leveling of electricity), [1].

not only reduces the mismatch between supply and demand but also improves the performance and reliability of energy systems and plays an important role in conserving the energy. Thermal energy can be stored during the un- peak period, usually at night, and re-use it during the peak load.

TES technology can reduce the generating and operating costs of cooling plant equipment. By utilizing TES, new generating plants can be eliminated. Moreover, some electric companies initiated different Tariff rate to reduce the use of electricity during the peak demands and force big consumers to store energy in the off-peak time by utilizing TES. In applications where peak loads occur only for a limited period during a year, such as worshipping places, which are used only for couple of hours during the day or the weak, TES systems can also be used, so that it can be to store the full need of cooling energy with reduced size equipment.

Thermal energy storage can be stored as a change in internal energy of a material as sensible heat, latent heat and thermochemical or combination of these. Cold storage technology has improved significantly since 1980 when electric utility companies recognized the need to reduce the peak demand on their generation and distribution systems. Chilled water, ice, or eutectic phase change materials are the cold storage media.

1.2 Classification of cold thermal storage

Cold thermal storage systems which are classified into thermal or chemical. The thermal CTS is classified into Sensible or latent heat storage system as shown in **Figure 2**.

1.2.1 Chilled water storage systems

Among all types of liquid water is selected to be the thermal storage medium, since it has the highest specific heat of all common materials (4.18 kJ/kg.°C). Chilled water with temperature 5 °C -7 °C can be generated by conventional chiller unites. The chilled water is then can be stored in an isolated concrete or stainless-steel tank for later use to meet the cooing needs, **Figure 3**. In general, in order to store 1 kWh of energy, a volume of approximately 0.1 m^3 is required. The chilled water can be pumped to the air handling unite (AHU) from the bottom of the tank and at the same time is substitute from the top by the return warm water from the AHU. Stable layers of water can be achieved due to variation of densities according on the temperature. This type of CTS is cost effective when the space is available. The chilled water tank by itself has other uses such as a back up water reservoir or for emergency fire extinguisher.

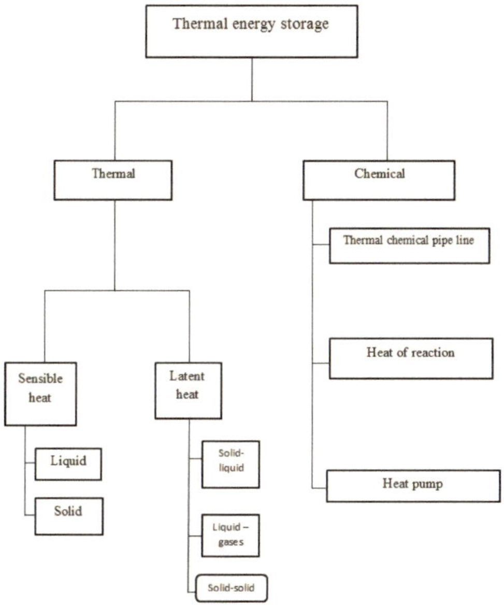

Figure 2.
Thermal energy storage types.

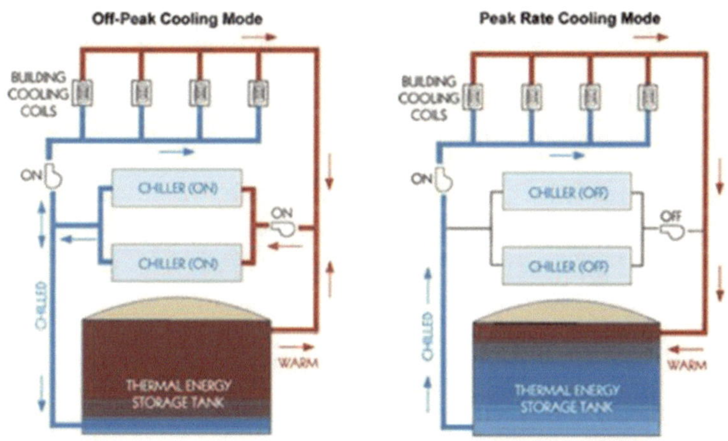

Figure 3.
Chilled water storage systems.

The quantity of stored heat in the storage tank can be calculated by

$$Q = \dot{m}C_p(T_e - T_i) \tag{1}$$

Where m is the mass of the chilled water,

1.2.2 Latent heat storage

When the materials go for a phase change from solid to liquid, liquid to gas or vice versa, it can store or release huge amount of heat which is latent heat. Latent heat storage (LHS) is getting more attractive because of the huge amount of energy that can be stored with small space which is usually one fourth less than the chilled water storage system. **Figure 4** shows the latent heat storage mechanism for solid liquid phase change.

Figure 5 shows a thermal storage system that utilizes ice. The size of the storage tank depends on the total volume of the melted liquid.

On the other hand, the ice storage system has COP of 2.5–4.1 which is less than that of chilled water storage system of 5–5.9 COP. So that the ice storage system economic benefit is beneficial for less Tarif at the off-peak time.

The storage capacity of the LHS system with a PCM medium is given by

$$Q = \int_{T_i}^{T_e} \dot{m} C_p dT + \dot{m} a_{fr} \Delta h + \int_{T_e}^{T_{fr}} \dot{m} C_p dT \tag{2}$$

Or:

$$Q = \dot{m}\left[C_p(T_e - T_i) + a_{fr}\Delta h + C_p\left(T_{fr} - T_e\right)\right] \tag{3}$$

The specific volumetric storage capacity of ice stores is 40–53 kWh m^3 [2].

Ice storage systems allow for innovative HVAC system design such as cold air distribution systems which have lower initial costs compared to conventional distribution systems. Ice storage systems include: Ice harvesters, Internal melt ice-on-coil storage systems, External melt ice-on-coil storage systems and Containerized ice storage systems. More details on each one of them can be found in Moncif Karaci "Energy Audit of Building System: An Engineering Approach".

1.2.3 Thermochemical energy storage

By a reversible chemical reaction, the energy can be absorbed or released for a thermochemical system. The stored thermochemical heat relies on the amount of storage material, the endothermic heat of reaction, and the extent of conversion.

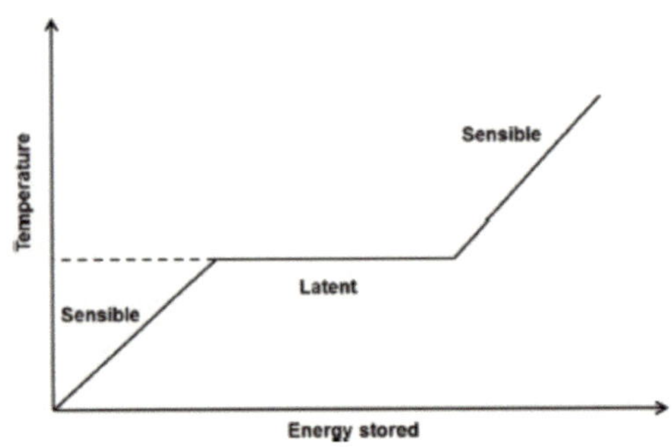

Figure 4.
Latent heat storage mechanism for solid –liquid.

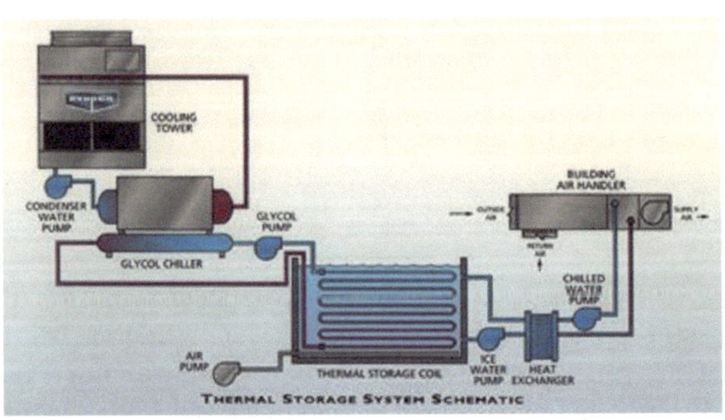

Figure 5.
Thermal storage system that utilizes ice, the size of the storage tank depends on the total volume of the melted liquid.

1.2.4 Eutectic systems

Salt and oil are two types of eutectic material. A eutectic salt can change its phase from liquid to solid at a specific temperature. These phases may have different crystal structures, or the same crystal structure with different lattice parameters. The phase change material such as Eutectic salts can have a freezing point of 8.5 °C which means less consumption of energy than ice. The PCM used in Eutectic system has less capacity than ice storage system but higher capacity than chilled water system. The eutectic system is more expensive and complex than chilled water systems and has warmest discharge temperatures (near 10 °C). More details of PCM materials can be found with [3].

1.3 Phase change materials

Phase change materials (PCM) are "Latent" heat storage materials. Unlike conventional (sensible) storage materials, PCM absorbs and release heat at a nearly constant temperature. They store 5–14 times more heat per unit volume than sensible storage materials such as water, masonry, or rock. PCM can be classified as organic, inorganic or eutectic which can be available in any required temperature range. The inorganic materials have almost double volumetric latent heat storage capacity (250–400 kg/dm3) than the organic materials (128–200 kg/dm3).

Organic materials are further described as paraffin and non-paraffins. Organic materials include congruent melting means melt and freeze repeatedly without phase segregation and consequent degradation of their latent heat of fusion, self-nucleation means they crystallize with little or no supercooling and usually non-corrosiveness. Inorganic materials are further classified as salt hydrate and metallics. These phase change materials do not supercool appreciably and their heats of fusion do not degrade with cycling.

The melting point of Paraffin varies between 5.5 °C and up to 75.9 °C, [4]. The latent heat of fusion varies from 170 kJ/kg and up to 269 kJ/kg. For the non-paraffin, the melting points also varies from 7.9 °C to 127.2 °C. They have heat of fusion ranging from 93 kJ/kg to 259 kJ/kg. The non-paraffin materials will have its own properties unlike the paraffin's, which have very similar properties. For the inorganic materials has a melting point varies from 16.7 to 102 and heat of fusion ranges from 146 to 242 kJ/kg. The high melting point allows les energy needed for those PCM to change state. The relatively high heat of fusion is another factor can affect the selection of such materials. **Table 1** shows the properties need to be considered in PCM. More details about the recommended PCM can be found in [4–6].

Properties	Features
Thermal	Melting temperature higher than or equal water freezing temperature, latent heat of fusion and heat transfer.
Physical	High density, small volume change, low vapor pressure.
Kinetic	No supercooling, adequate crystallization rate
Chemical	Stability, compatibility, toxity, non-flammable
Economics	Availability and cost effective

Table 1.
Required features for the PCM materials.

2. Charging of cold thermal storage

2.1 Fundamental studies related to cold thermal storage

Ice on tube system essentially consists of cold pipes submersed in stagnant water or cold pipes submersed in a cross flow of water. Freezing of water was studied by several researchers. **Table 2** shows a summary of the works that have been done by several researchers to study the freezing phenomena as well as melting.

Type of study	References	Finding
Freezing of water on immersed tubes bank	[7, 8]	Ice layers affect negatively heat transfer
solidification of PCM around a cylinder for ice-bank applications	[9, 10]	The inlet temperature influences the freezing process
ice formation around a horizontal long copper tube U-shap	[11]	Slope of the ice thick-ness, in which the axial distance depended on time but varied with coolant flow rate and Stanton and Biot numbers
solidification of PCM around a cylinder for an ice-bank application	[9, 10]	Lower initial temperature of the liquid phase seemed to accelerate the solidification.
Additive of stainless steel pieces, copper pieces and PCM-graphite composite material to the PCM	[12, 13]	Addition of stainless steel pieces in the PCM does not increase the heat flux significantly. However, addition of copper pieces and the use of graphite composite enhance heat transfer significantly.
Ice storage system using water–oil mixture	[14]	Slush ice is formed out of the tube surface
Analyzed ice formation around a finned-tube	[15, 16]	The finned enhanced the ice formation by 45%
Finned tubes to enhance heat transfer for forming of ice on tubes in bank geometry, **Figure 6**	[17]	Fins diameter and temperature of the coolant enhance the freezing process
Thin ring and annular fin enhancing ice formation	[18]	Thin rings have better performance in comparison to annular fins
Spray chemicals to enhance the defrosting process	[19]	Ethanol inhibits of the frost layer and on the same time provide better defrosting effect
Mathematical model for the optimization of ice making in a laminar falling film on vertical plates	[20]	The freezing time can be reduced by six times for commercial model
Studied the freezing of a falling film on horizontal tubes	[21, 22]	They showed that ice formation depends on falling film and coolant flow rates. Also, the overall heat transfer coefficients are controlled by ice thermal resistance
Investigate the impacts of shape-stabilized phase change material (SSPCM) and different control strategies on the energy consumption and peak load demand as well as electricity cost	[23]	Use of SSPCM in the building could reduce the building electricity cost significantly. However, the use of ice systems is more attractive than that of chilled water systems
Economic opportunities afforded by installing an ice storage system to existing air conditioning	[1]	Combining both the energy storage and an incentive time structured rate shows reasonable daily bill savings

Table 2.
Summary of the works that have been done by several researchers to study the freezing phenomena as well as melting.

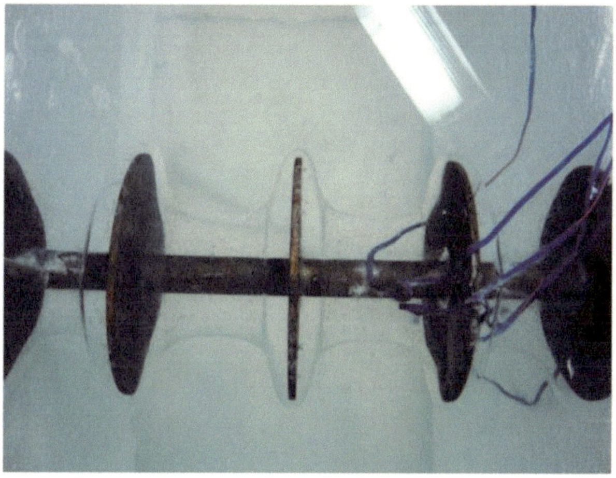

Figure 6.
Ice accumulation on finned tube, [17].

2.1.1 Falling film phenomena

Falling film can be either on vertical flat plate or horizontal tubes to maximize the heat transfer coefficient. Three main modes were shown when the liquid drop from any tube. It can take a shape of droplet, jet or sheet depending on the flow rate as shown in **Figure 7**. Combined modes usually took place. By increasing the flow rate, the mode changes to droplet-jet, jet, jet- sheet and finally sheet mode. Multi-modes can be appeared at the same time for the same flow rate, depending on the number of tubes aligned vertically. More secription of the flow pattern for the falling film can be found in [24–27]. For the purpose of freezing for the CTS, several articles talked about the possibility of utilizing the advantages of high heat transfer coefficient to be used for freezing.

For the maximum falling film flow, the sheet mode took place. This mode is shown clearly in the upper two test tubes, the rest shows sheet-jet mode. Some of the falling film splashed from the down tubes. By reducing the flow rate, jet mode is achieved and it is observed clearly on the upper tubes and the rest is a jet-droplet mode. A small amount of the falling film splashed from the down tubes. With further decrease in the flow rate, a steady droplet mode all over the test tubes was noticed with no splash from the tubes.

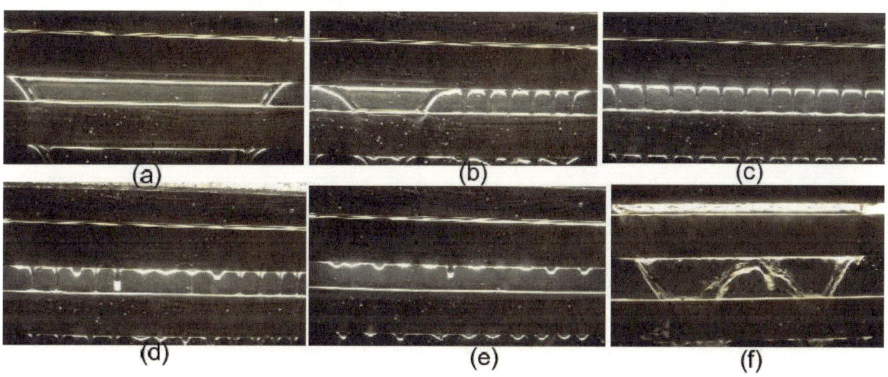

Figure 7.
Idealized innertubes falling film modes: (a) sheet mode, (b) mixed of sheet and jet, (c) jet mode, (d) droplet-jet mode, (e) droplet mode, (f) sheet mode [24].

2.2 Ice formation characteristics

The formation of ice begins usually where the inlet coolant exists due to the lowest temperature there. The regularity of ice depends on the types of stream that moves on the tubes. It is obvious that the formation of ice increases with time. However, the ice formation reduces heat transfer due to an increase in the insulation of ice (increased thickness of ice causes an increase in thermal resistance and consequently reduction of heat transfer through ice layer). It has been observed that as the ice accumulates on the test tubes, it takes longer time to remove it from the test tubes by the discharge cycle and relatively large quantity of ice is melted.

Ice accumulates circumferentially on the tube surfaces until the tube spacing is filled with ice which makes it harder in discharging process (**Figure 8**).

2.3 Heat transfer analysis for freezing

Heat transfer assessment on tubes bundle is based on detecting of temperature change, temperature of the liquid flowing through the tube bundle and temperature of outside liquid with simultaneous measuring of the current flow of liquids. This way, the real value of the heat transfer is determined at the tube bundle working under the set conditions. And subsequently, the corresponding heat transfer coefficient can be determined. Series arrangement has the advantage of high flow rate inside the tube and consequently high Reynolds number and high inside heat transfer coefficient. However, it difficult to be designed and maintain. In contrast, parallel arrangement is more convenient for industry, but reduces the inside heat transfer coefficient due to the inside flow is divided in to the all tubes and hence reducing the Reynolds number and so the inside heat transfer coefficient.

For the CTS, the heat transferred from the outside liquid to the entire coolant. The amount of heat transferred is determined from the temperature difference and the corresponding fluid flow rate. When solidification is taken place, the outside liquid loss sensible heat to reach the freezing point and then latent heat, to form ice. The rest of the outside liquid cools down while it is contacting the outside of the cold tubes.

The heat transfers from the outside liquid to the inside coolant via ice layer and the tube wall as shown in **Figure 9**. The thermal conductivity of the tube wall is usually very high comparing to the ice layer, so the tube resistance is usually ignored.

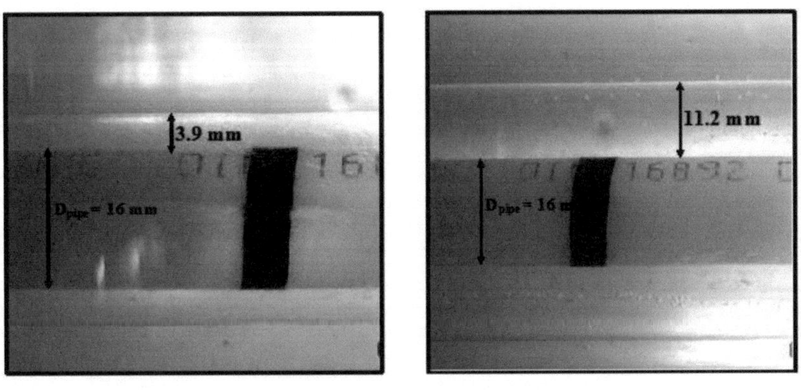

Figure 8.
Ice photo on the formed ice on the tube at time of (a) 50, (b) 198 minutes.

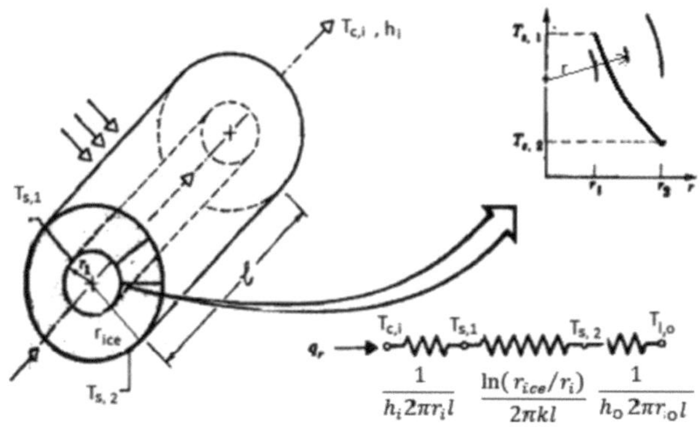

Figure 9.
Thermal resistance for ice accumulates on the tubes [28].

The rate of heat transfer from the outside fluid to the coolant inside the tube can be calculated as:

$$\acute{Q}_{ins} = \frac{(T_{\infty,o} - T_{ave,i})}{R_o + R_{ice} + R_i} \tag{4}$$

For thermal resistance shown in Fig., then

$$\acute{Q}_{o-i} = \frac{(T_{\infty,o} - T_{ave,i})}{\frac{1}{h_o 2\pi r_{ice} l} + ln} \tag{5}$$

2.3.1 Heat transfer coefficient inside the tube h_i

The inside heat transfer coefficient is given by:

$$h_i = NuK/D \tag{6}$$

Where:

In order to get *Nu, Re* must be calculated and is given by, $\mathfrak{R} = \rho u_m D_h/\mu$,where the velocity of the coolant at the test tube, $u_m = \acute{m}_c/\rho A_i$ and, $\acute{m}_c$ is mass flow rate of the coolant which enters to the test section.

For constant surface temperature condition, and laminar flow, Re ≤ 2300, Nusselt number is given by many researchers. One is given by Incropera, F.P. et al., as:

$$Nu = 1.86\left[\frac{RePr}{l/D}\right]^{1/3} + \left[\frac{\mu}{\mu_s}\right]^{0.14} \tag{7}$$

Where this equation can be applied for the following condition:
T_s (surface temperature) = constant.

$$0.48 < \text{Pr} < 16{,}700.$$

$$0.0044 < (\mu/\mu_s) < 9.75.$$

And for turbulent flow, where Re ≥ 10,000, Nusselt number can be calculated by

$$Nu = 0.023\mathfrak{R}^{0.8}Pr^{0.3} \tag{8}$$

Where the ranges are.

$$0.7 \leq \Pr \leq 160.$$

$$l/D \geq 10.$$

All properties except μ_s should be evaluated at the average value of the inlet and outlet temperature, $T_{ave} = (T_i + T_o)/2$.

2.3.2 Outside heat transfer coefficient

[29, 30], developed correlations to approximate the heat transfer coefficient for the outside flow (falling film), As follows:

For high Re, (sheet mode):

$$Nu = 2.194\Re_f^{0.28} Pr^{0.14} Ar^{-0.2} \left(\frac{s}{d}\right)^{0.07} \tag{9}$$

For medium Re, (jet mode):

$$Nu = 1.378\Re_f^{0.42} Pr^{0.26} Ar^{-0.23} \left(\frac{s}{d}\right)^{0.08} \tag{10}$$

For low Re, (droplet mode):

$$Nu = 0.113\Re_f^{0.85} Pr^{0.85} Ar^{-0.27} \left(\frac{s}{d}\right)^{0.04} \tag{11}$$

The liquid properties were evaluated at film temperature T_f.

Where.

Nu is modified Nusselt number $\left(\frac{\nu^2}{g}\right)^{1/3} h/k$.

Re_f is film Reynolds number $2\Gamma/\mu$ Pr

Pr is Prandtl number $C_p\mu/k$ (1.20)

Ar Archimedes number based on tube diameter $d^3 g/\nu^2$

Eq. 4 estimate the instantaneous heat transfer ac certain time for specific ice radius. For the whole experiment run for specific time, we need to get the total heat absorbed from the outside fluid. The outside fluid is either at stationary state or moving across the tube. For the first case if the outside liquid at constant condition

$$\acute{Q}_{s,l,c} = \frac{m_l}{t} C_{p,l}(T_{li} - T_{lf}) \tag{12}$$

Where T_{li} and T_{lf} are the initial and final temperature of the outside liquid respectively.

For the second case where the outside fluid is moving across the tubes, the sensible transfer heat rate is calculated by

$$\acute{Q}_{s,l,m} = \acute{m}_l C_{p,l}(T_{li} - T_{lo}) \tag{13}$$

Where T_{li} and T_{lo} are inlet and exit temperature of the outside liquid.

Part of the outside liquid freezes on the outside tube. The freezing consumes latent heat which can be calculated as:

$$\acute{Q}_{fr} = \frac{M_{ice} L}{frozentime} \tag{14}$$

Where L is the latent heat of fusion for the PCM.

The formed ice is now cooled to a temperature below the freezing temperature which is consider as sensible heat and is given by

$$\acute{Q}_{s,s} = M_{ice}C_{p,ice}\left(T_f - T_c\right)/time \tag{15}$$

Hence the total heat transfer through the ice layer can be computed as:

$$\acute{Q}_{O-i} = \acute{Q}_{s,l} + \acute{Q}_{fr} + \acute{Q}_{s,s} \tag{16}$$

The total heat transfer absorbed by the coolant can be given by:

$$\acute{Q}_{c,act} = \acute{m}_c C_{pc}(T_{co} - T_{ci}) \tag{17}$$

where the coolant flow rate $\acute{m}_c$ (kg/s), and $T_{ci} \wedge T_{co}$ are its inlet and outlet temperature.

The average rate of heat transfer from the test tubes to the coolant at a specific time, $\acute{Q}_{c,ave}$ is calculated by the following equation:

$$\dot{Q}_{c,ave} = \left[\int_{t=0}^{t=n} \dot{Q}_c \, dt\right]/time \tag{18}$$

where:
t: is the interval time in minutes
n: is the indicated time
This integration can be obtained by the trapezoidal rule of integration:

$$\int_a^b f(x)dx = \frac{b-a}{2n}\left[f(x_0) + 2f(x_1) + 2f(x_2) + \dots + 2f(x_{n-1}) + f(x_n)\right] \tag{19}$$

By substituting the h_i and h_o in Eq. 10 to get the overall heat transfer coefficient and then substitute in Eq. 9 to get the calculated rate of heat transfer from the test tubes to the coolant at a specific ice thickness, $\acute{Q}_{c,cal}$ which needs to be compared with the total heat transfer to the coolant $\acute{Q}_{c,act}$ which was given from Eq. 6.

The experimental overall heat transfer coefficient, h_ov is determined using the equation

$$h_{ov} = \frac{\acute{Q}_{,o-i}}{A_{ice}\Delta T_{lm}} \tag{20}$$

where, A_{ice} is the heat transfer area and ΔT_{lm} is the logarithmic temperature difference between the working fluid flows. The ice layer affects the heat transfer coefficient since the thermal conductivity of the ice is low and as its thickness increases the thermal resistance increases which lead to lower heat transfer coefficient.

2.4 Affecting parameters on freezing

Ice freezing quantity is affected by the outside flow rate (mode of the falling film) and its temperature, as well as by coolant flow rate and coolant temperature (**Table 3**).

2.4.1 Effects of the falling film behavior on ice accumulation

In the beginning of the freezing process high heat transfer coefficient was shown which reaches to 1000 W/m^2.K, which shows the influence of the failing film.

Parameter	Effect
Effect of time on accumulated mass of ice, **Figure 10**	Ice formation distinctly increases with time. As ice thickness increases, its thermal resistance increases and its heat transfer decreases. Because ice has a low thermal conductivity, it acts as an insulator for the heat transfer.
Effect of coolant flow rate on accumulated mass of ice, **Figure 11**	The inner heat transfer coefficient due to increase in Reynolds number and hence increases the inner heat transfer coefficient.
Effect of outer flow on accumulated mass of ice, **Figures 11** and **12** and	The outer heat transfer coefficient due to increase in Reynolds number and hence increases the outer heat transfer coefficient.
Coolant temperature, **Figure 13**	More ice accumulated on the tube
Tube diameter	Big effect, since the area is increased
Tube spacing	Low effect
Tubes arrangements Effects, **Figures 14** and **15**	Affect Ice formation. Because of flow rate and the temperature difference and the temperature variation along the tubes.
Adding Fin, **Figure 16**	Enhance the ice accumulation

Table 3.
Effects of different parameters on the performance of cold thermal storage.

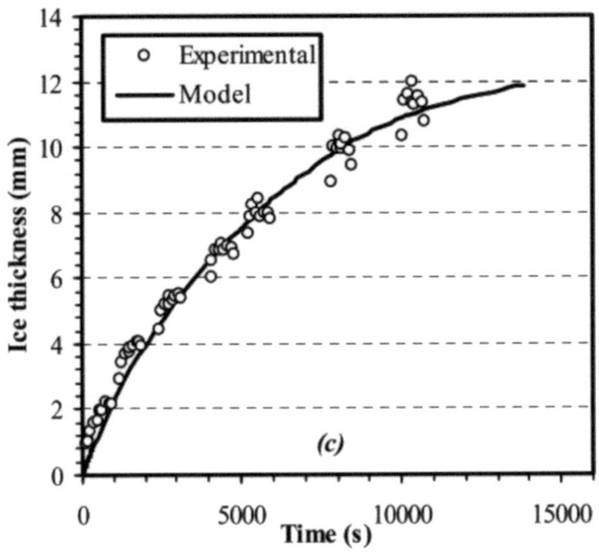

Figure 10.
The effect of ice accumulation with time on the rate of freezing.

However, by the further accumulation of ice, the heat transfer coefficient drops as of the effect of low thermal conductivity of ice. To get the best advantages of the falling film an optimum design must be applied to have quick charging and discharging operations so, more ice can be formed in short time and then collect it at the bottom of the reservoir. The falling film quantity must be adequate and the flow rate must handle the drag force of the tubes.

2.4.2 Falling film backsplash

While the falling film falls over the tubes vertically, some quantity leaves the falling film stream and falls outside the stream and drops to the bottom of the reservoir without further collision with the rest of the tubes. This phenomenon is

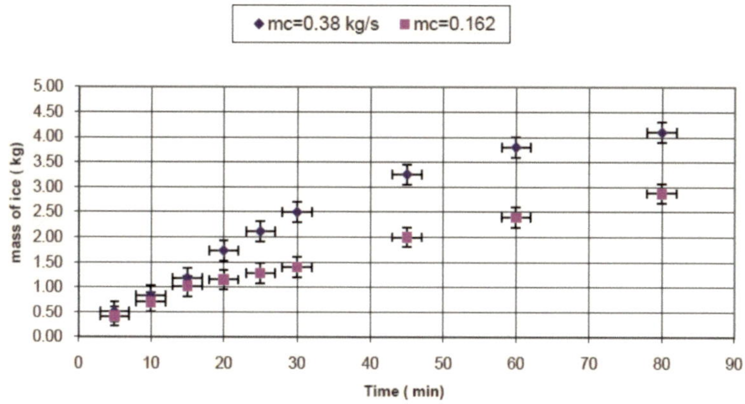

Figure 11.
Comparison of the formed ice for different coolant flow rate.

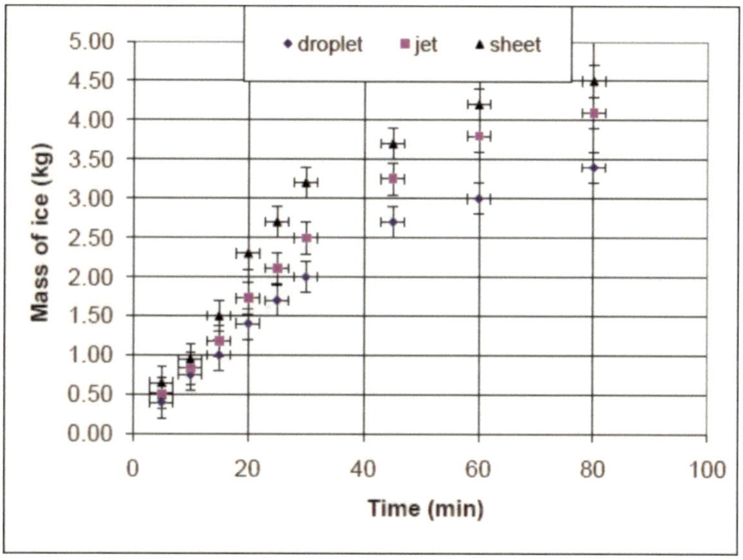

Figure 12.
Comparison of the formed for different falling film flow rate.

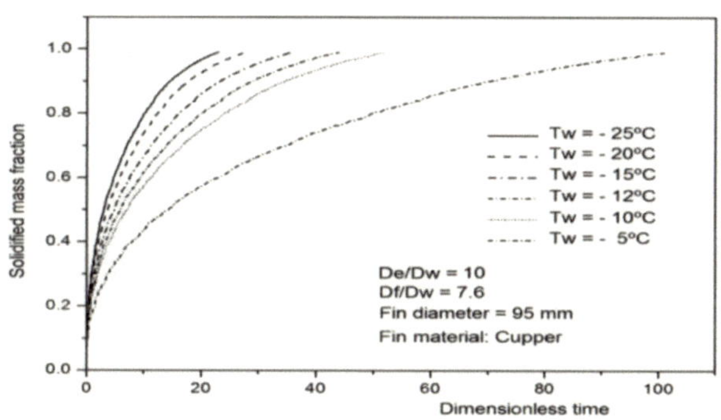

Figure 13.
Effect of variation of cooling temperature on freezing, [17].

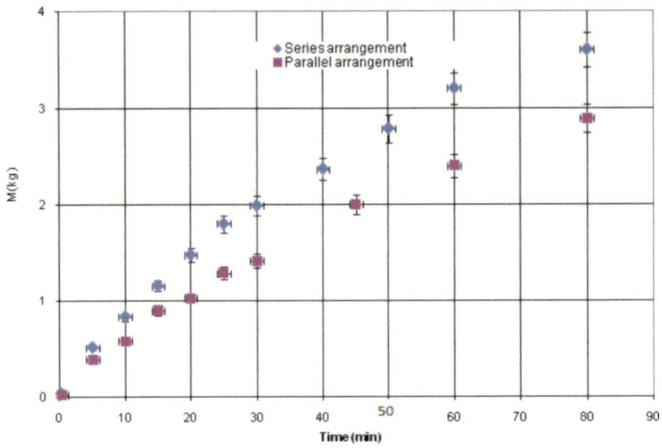

Figure 14.
Comparison of the formed ice between series and parallel arrangements for $\dot{m}_c$ =0.162 kg/s & $\dot{m}_l$ =0.025 kg/s.

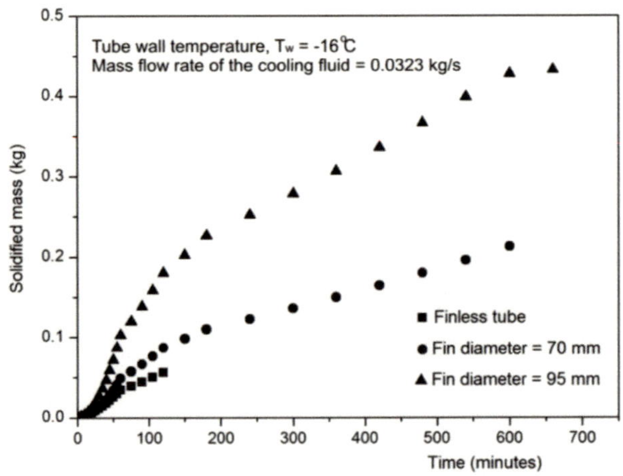

Figure 15.
Comparing of ice quantity for tube with fin and finless tube, [17].

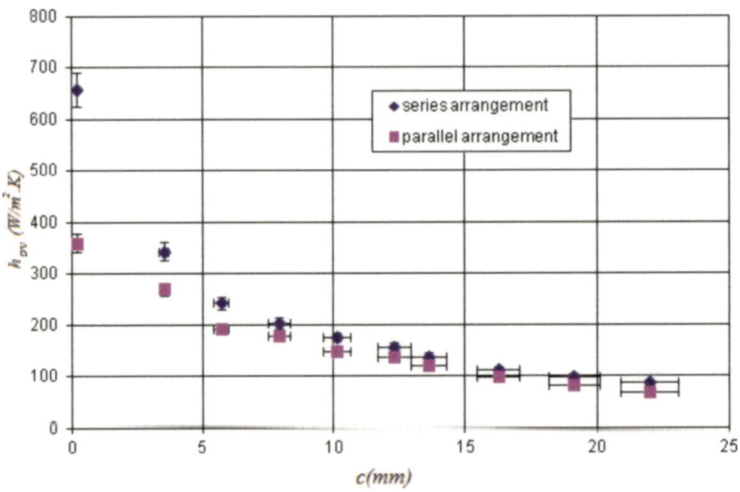

Figure 16.
Comparison of the experimental overall heat transfer coefficient between series and parallel arrangements for $\dot{m}_c$ = 0.162 kg/s & $\dot{m}_l$ =0.025 kg/s.

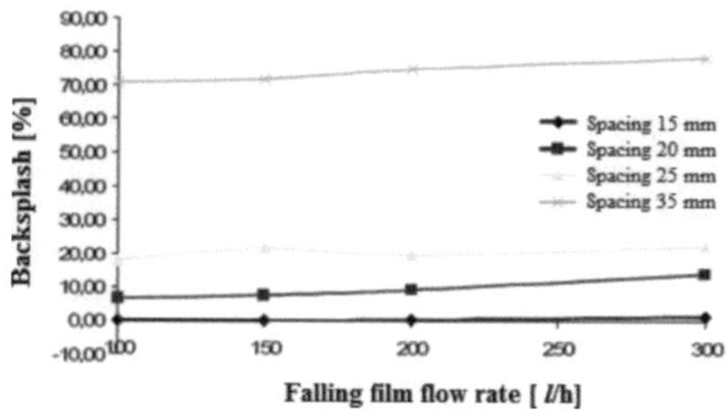

Figure 17.
Backsplash as shown in [31].

called "backsplash". The result is shown in **Figure 17** was created by Jiri Pospisil, et al. The backsplash increases with increasing of tube spacing and can reach up to 70% of the total flow falls over the test section for tube spacing of 35 mm. The backsplash also increases with increasing of the falling film flow rate but with only small percentage.

3. Discharging of ice and reusing it

In CTS systems the ice can be stored either on the tubes or in the isolated reservoir. For the ice on the tube system, the discharge system can be either inside the tube or outside the tubes. For the collected ice in an isolated reservoir, the ice must first be released from the tubes and then passing the warm water through the accumulated ice. **Table 4** provide previous effort for the discharging system to achieve the maximum capacity out of the CTS system.

Type of study	Reference	Finding
The study was on spherical enclosures	[32]	Shows an efficiency of 90% for extracting the cold out of the CTS
The study was about cooled cylinders, that are arranged in staggered and in line.	[33]	The discharge time was about half of the freezing time.
Model the internal melting for the ice on tube	[34]	The model considers the variation in temperature and discharge rate of discharge
Study the discharge on heated horizontal cylinders	[35]	Most of the melting time occurs when the tube is surrounded by ice
Modeling the discharging process for coil pipes using n-Tetradecane as a phase change material	[36]	Less time is needed for higher flow rate and inlet temperature. No big effect for the variations of the tube diameter
Study liquid–ice thermal storage system	[37]	Less time is needed for discharging, than that of charging time
Slush ice in a horizontal cylindrical capsule is studied	[38]	The melting rate increases as a function of melting time
Crystal ice formation of the solution and its removal phenomena at a cooled horizontal solid surface	[39]	Heat flux and inclination enhance the ice removal

Table 4.
Summary of the works that have been done by several researchers to study the melting phenomena.

3.1 Heat transfer analysis of discharging

The absorbed heat required to release ice consists of the following heats: sensible heat of sub-cooled ice, latent heat of melted ice and sensible heat of melted water. Thus, the experimental ice melting be expressed as

$$Qexim = Q\ smice + Q\ Lmice + Q\ sw \tag{21}$$

where

$$Qsmice = M_{mice} C_{p,mice}\ (Ts{-}T_O)/\tau_{mice} \tag{22}$$

and

$$QL_{mice} = M_{mice} L_{mice}/\tau_{mice} \tag{23}$$

and

$$Qsw = M_{mice} C_{p,w}(T_w{-}T_O\ )/\tau_{mice} \tag{24}$$

This heat is added by the heating solution Q_{hav} which is expressed as

$$Q_{hav} = \frac{\int_{t=0}^{t=n} \mathrm{Qhdt}}{\mathrm{time}} \tag{25}$$

where
t is the time interval, n is the indicated time, and

$$Q_h = \dot{m}_h\ C_{p,h}(T_{hi} - T_{ho}) \tag{26}$$

The overall heat transfer coefficient U_{exim} of the experimental ice releasing is expressed as follows:

$$U_{exim} = Q\ exim/[A_{ice}\ (\bar{T}_h{-}T_O)] \tag{27}$$

The fluid properties of water and ice are listed in **Table 5**.

3.2 Characteristics of ice releasing

The discharge cycle can melt the ice from inside to outside or from outside to inside depending on the flow of the warm brine that flow to the chiller. For the first case, the ice begins to melt from the inner surface of the formed ice toward the outer surface forming liquid between the tube surface and the ice layer. The ice layer is then decreases to the point that, it falls and collected at the bottom of the

Fluid	μ, N s m-2	ρ, kg m-3	k, W m-1 K-1	C_p, kJ kg-1 K-1
Water (0.25 °C)	1750x10–6	1000	569.6 x10–3	4.2
Ice (−5 °C)	—	920	1.88	2.040
Cold solution (−10 °C)	8 x10–3	1060	0.467	3.3488
Hot solution (25 °C)	3.37 x10–3	1050	0.43	3.30

Table 5.
Fluids properties.

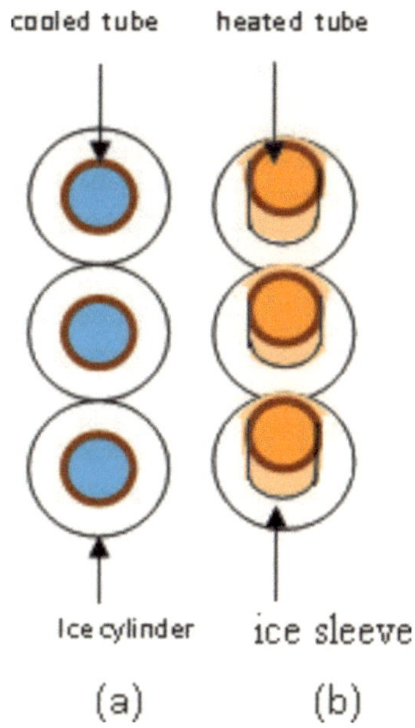

Figure 18.
Schematic for (a) ice formation and (b) ice releasing.

reservoir. **Figure 18** displays a schematic diagram for ice formation and releasing. For the second case where the warm water falls on the outside of the ice layers, the ice layer has more time to tick to the tube. In charging cycle, it is mandatory not allowing the ice layer between tubes to stick together, otherwise it will take much more time to get released.

3.2.1 Released ice percentage

During discharge process, some ice is melted and others falls to the bottom of the reservoir. **Figure 19** shows the percentage of the released solid ice for various average ice thicknesses for an experiment done by [10]. The temperature of the internal solution (the brine) is also affected the melting process as shown in the

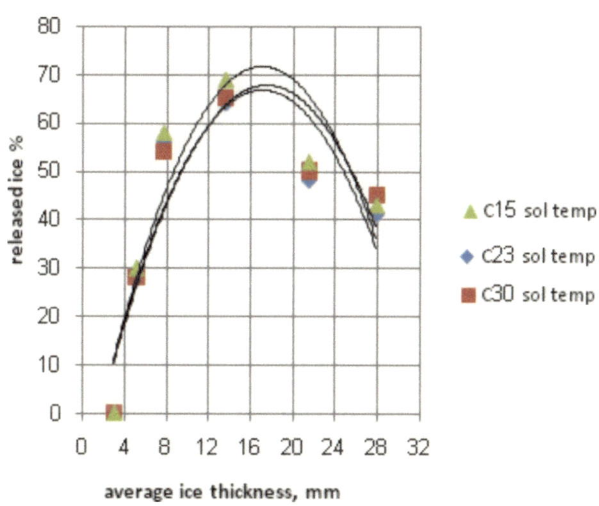

Figure 19.
Effect of heating solution temperature on released ice percentage.

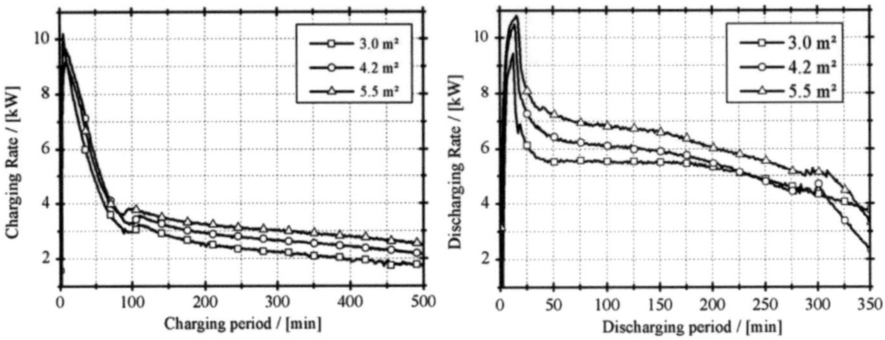

Figure 20.
Charging and the discharging rate for a 10 kW CTS system, was applied in the Institute for Thermodynamics and Thermal Engineering (ITW) of the University of Stuttgart [2].

Figure which shows a small effect on the percentage of ice releasing. Koller et al. [2] studied charging and discharging for CTS and found that the discharging is about 70% benefit of the charging system as shown in **Figure 20**.

3.3 Heat transfer coefficient for ice releasing

The experimental overall heat transfer coefficient depends on heat transfer from the heated solutions to the solid ice which consists of sensible heat for sub-cooled ice, latent heat of melted ice and sensible heat of melted water. The ice surface area and the temperature difference between the heating fluids and melting point affects the experimental overall heat transfer coefficient U_{exim}. **Figure 21** shows the variation of the experimental overall heat transfer coefficient U_{exim} for solid ice releasing

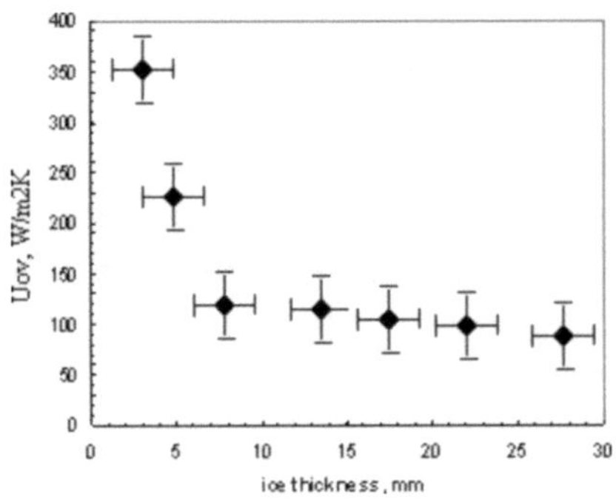

Figure 21.
Effect of ice thickness on U_{ov}.

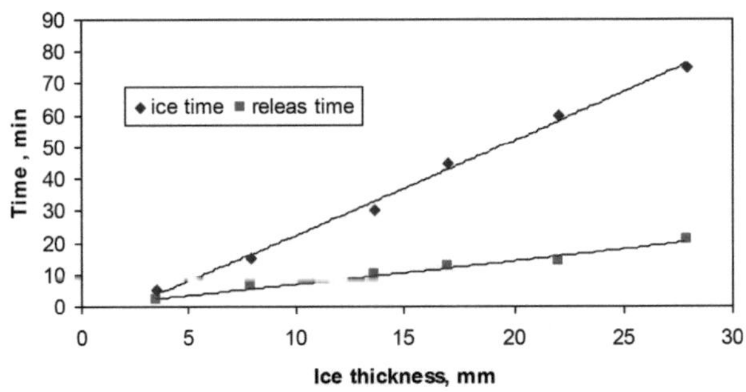

Figure 22.
Effect of ice thickness on the time of ice formation and ice releasing.

with ice thickness for the experiment done by [40]. As the ice thickness decreases as melting is taken place, U_{exim} is increasing to reach its maximum value of 350 W/m^2 K when the ice thickness decreases to 4 mm. Usually the releasing of ice time is shorter than the freezing time, which may varies from 1/2 to ¼ depending on the ice thickness and the warming solution used for discharging (**Figure 22**).

4. Enhancing both charging and discharging for cold thermal storage

4.1 Utilize thin film

Grooved tubes can be used which increases the heat transfer surface area and hence enhancing the heat transfer coefficient. The grove can also create a thin film which enhance the heat transfer as mention by [41]. Utilizing such techniques can reaches a heat flux of 1.400 MW/m^2 as shown in **Figure 23**.

4.1.1 Using extended surface on the tube

The cheat transfer can be enhanced on the cold tube by having extended surface. This will help not only in enhancing the heat transfer, but in also discharging the ice out of the tube, in cases when requires many on and off cycles to collect ice.

4.1.2 Using suitable PCMs

PCMs have features of higher melting points than ice and reasonable latent heat of fusion which allow them to require less energy during charging cycle. The economic wise play an important role in utilizing such materials in PCMs

4.1.2.1 Cost effectives of energy storage

Thermo-economic optimization of an CTS system should be carried out, which considers the environmental aspects, and cost effective. The cost includes the capital and operational costs as well as the penalty cost due to CO_2 emission. The pay back periods will be a major influence to convince investors to utilize the CTS. Electric Power Research Institute in USA have put a methodology for energy storage Valuation, shown in **Figure 24**. The methodology consists of four steps. It starts

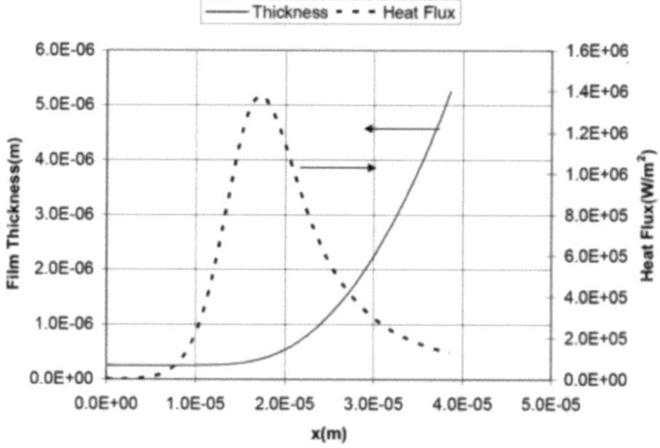

Figure 23.
Thin film evaporation scenario.

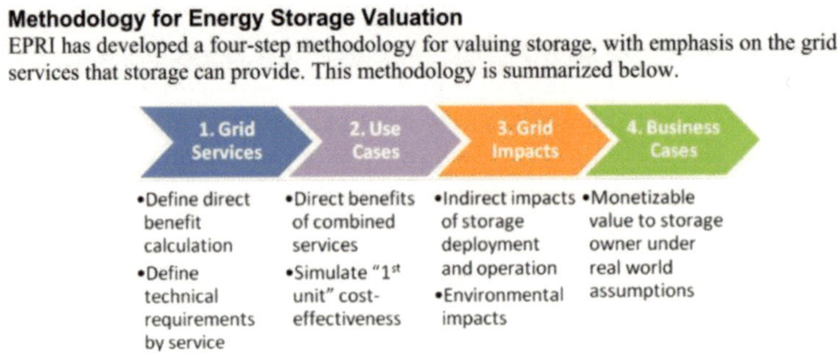

Figure 24.
Overview of electric power research institute (EPRI) energy storage valuation methodology.

by defining the expected direct benefits of the CTS and the technical requirements. It then required to simulate the unite cold effectiveness. It then need to define the indirect impact such as the environmental impact. The last step is to study the business issue, whether can this utilization of CTS convert into a cash or not. Lot of works were conducted for the feasibility study of utilizing the CTSs like [1, 42].

5. Summary

1. The rate of the ice formation depends on inlet and outlet flow rate.

2. The heat transfer coefficient is affected negatively by accumulation of ice on the tubes.

3. The tube arrangement has a large effect on ice formation and needs to be considered.

4. Falling film phenomenon can benefit the charging and discharging process and need to be optimized.

5. The heating solution temperatures have a small effect on the gained ice and releasing time.

6. The heat transfer coefficient is affected by the direct contact area between the ice sleeves and the heated tubes.

7. Cost effivenss should be conducted to study the feasibility of adopting the CTSs.

This chapter helps in the analyzing and constructing of cold thermal storage systems, which are necessary for energy saving today.

Nomenclature

a_{fr}	fraction frozen
A	area, m^2
C_p	specific heat, kJ/kg.K
D	diameter, m^2

Δh	heat of fusion J/kg
L	latent heat of fusion
ℓ	length of the test tubes, m
m	mass, kg
M	mass of ice, kg
•*m*	mass flow rate, kg/s
Q	quantity of stored heat, J
$\acute{Q}$	rate of heat transfer, W
T	temperature, °C

Greek Symbols

ν	dynamic viscosity. N.s/m2
ρ	density, kg/m3
ε	The test section heat exchanger effectiveness

Subscripts

a,f	absorbed from the falling film
ave.	average
c	coolant
e	end
f	film
fr	freezing

Author details

Hani Hussain Sait
King Abdulaziz University, Rabigh, Saudi Arabia

*Address all correspondence to: hhsait@kau.edu.sa

References

[1] Badr Habeebullah, Rahim Jassim, Nedim Turkmen, Ahmad Bokhary, Majed Alhazmy. Cost-Effectiveness Opportunities for Thermal Energy Storage Systems: A Case Study of School Building in Saudi Arabia. International Journal of Sustainable and Green Energy. Vol. 5, No. 4, 2016, pp. 59–70.

[2] Koller, T., K. Spindler, and H. Müller-Steinhagen, Experimental and theoretical investigations of solidification and melting of ice for the design and operation of an ice store. International Journal of Refrigeration, 2012. 35(5): p. 1253–1265.

[3] Ahmet Fertelli, Orhan Buukalaca and Alper Yilmaz, "Ice Formation Around a Horizontal Tube in a Rectangular Vessel" *Journal of Thermal* Science and Technology, 2009; 29(2): 75–87

[4] Sharma, A., et al., *Review on thermal energy storage with phase change materials and applications.* Renewable and Sustainable Energy Reviews, 2009. **13** (2): p. 318-345.

[5] Fang, G., et al., Preparation and characterization of nano-encapsulated n-tetradecane as phase change material for thermal energy storage. Chemical Engineering Journal, 2009. 153(1): p. 217–221.

[6] Oró, E., et al., Review on phase change materials (PCMs) for cold thermal energy storage applications. Applied Energy, 2012. 99: p. 513–533.

[7] Intemann P.A. and Kazmierczak M., "Heat Transfer and Ice Formation Deposited upon Cold Tube Bundles Immersed in Flowing Water, Convection Analysis". Int J of Heat Fluid Flow; 1997 40 (3) ;557–572.

[8] Intemann P.A. and Kazmierczak M.1994, "Convective heat transfer for cold tube bundles with ice formations in a stream of water at steady state". Int. J of Heat Fluid Flow 1994; (15) :491–500.

[9] Ismail, K. A.R, Jesus, A. B. de, 2001a, Parametric study of solidification of PCM around a cylinder for ice-bank application, Int. J. Refrig. 24, 809–822.

[10] Ismail, K.A.R. and A.B. de Jesus, Parametric study of solidification of PCM around a cylinder for ice-bank applications. International Journal of Refrigeration, 2001b. 24(8): p. 809–822.

[11] Habeebullah, B. A., "An experimental study on ice formation around horizontal long tubes" Int J Refrig, 2007; 30(5): 789–797

[12] Cabeza L. F., Mehling H., Hibler S., Ziegler F., 2002a. Heat transfer enhancement in water when used as PMC in thermal energy storage, Appl. Therm. Eng., 22, 1141–1151.

[13] Cabeza, L.F., H. Mehling, S. Hiebler, F. Ziegler, "Heat transfer enhancement in water when used as PCM in thermal energy storage" Appl Therm Eng, 2002b; 22(10):1141–1151

[14] Matsumoto, K., et al., *Ice storage system with water–oil mixture: formation of suspension with high IPF.* International Journal of Refrigeration, 2000. **23**(5): p. 336-344.

[15] Kayansayan N. and Acar M. A., "Ice formation around a finned-tube heat exchanger for cold thermal energy storage" International journal of thermal science 45 pp. 405–418, 2005a.

[16] Kayansayan, N., Acar, M. A., 2005b, Ice formation around a finned-tube heat exchanger for cold thermal energy storage, INT J THERM SCI 45, 405–418

[17] Ismail, K., P. Silva, and F. Lino, Enhancement of solidification of PCM

around finned tubes: modeling and validation. Vol. 06. 2016. 49–63.

[18] Jannesari, H. and N. Abdollahi, Experimental and numerical study of thin ring and annular fin effects on improving the ice formation in ice-on-coil thermal storage systems. Applied Energy, 2017. 189: p. 369–384.

[19] Jia, Y., et al., Experimental studies on frost and defrost of fine tube bundles under coolant temperature between −20 and −58°C. International Journal of Heat and Mass Transfer, 2018. 116: p. 617–620.

[20] Cliché, A. and Marcel Lacroix, 2006 "Optimization of ice making in laminar falling films", Energy Convers Manage, Vol. 47; 2260–2270.

[21] Sait H. H., Hussain A. and Selim A. M., "Experimental investigation on freezing of water falling film on vertical bank of horizontal cold tubes" *J. Thermal Sci. Eng. Appl.* 2012 ; 4: 041006–1- 041006-7

[22] Sait, H.H., Heat transfer analysis and effects of feeding tubes arrangement, falling film behavior and backsplash on ice formation around horizontal tubes bundles. Energy Conversion and Management, 2013. 73 (Supplement C): p. 317–328.

[23] Zhu N, Wang S, Ma Z, Sun Y. "Energy performance and optimal control of air-conditioned buildings with envelopes enhanced by phase change materials". Energy Convers Manage 2011; 52: 3197–3205

[24] Wang, X., et al., Flow Patterns and Mode Transitions for Falling Film on Flat Tubes, in International Refrigeration and Air Conditioning Conference. 2010.

[25] Austegard, A., et al., Flow Pattern Transitions in and Hysteresis Effects of Falling Film Flow over Horizontal Tubes Related to LNG Heat Exchangers. Energy Procedia, 2015. 64: p. 23–32.

[26] Roques JF. And Thome JR., "Falling Film Transitions between Droplet, Column and Sheet Modes on a Vertical Array of Horizontal 19 fpi and 40 fpi Low Finned Tubes". *Heat Transfer Engineering*; 2003a:24(6); 40–45

[27] Roques, J.-F. and J.R. Thome, Falling Film Transitions Between Droplet, Column, and Sheet Flow Modes on a Vertical Array of Horizontal 19 FPI and 40 FPI Low-Finned Tubes. Heat Transfer Engineering, 2003b. 24(6): p. 40–45.

[28] Incropera, F.P and D.P. DeWitt. Introduction to Heat Transfer, Wiley; 4th edition 2001.

[29] Hu X. and Jacobi A.M., "The Intertube Falling Film: Part 1-Flow characteristics, Mode Transitions and Hysteresis". Journal of Heat Transfer; 1996a ; 118: 616–625.

[30] Hu X. and Jacobi A.M., "The Intertube Falling Film: Part 2 Mode Effects on Sensible Heat Transfer to a falling Liquid film". Journal of Heat Transfer; 1996b; 118: 626–633.

[31] Jiri Pospisil, Libor Chroboczek, Zdenek Fortelny and Pavel Charvat, "Falling film heat exchange and backsplash on horizontal tube bundles" *International Journal Of Energy*, 2009; 3(3), 35–42

[32] Eames I. W., Adref, K. T., 2002. Freezing and melting of water in spherical enclosures of the type used in thermal (ice) storage system, Appl. Therm. Eng. 22, 733–745.

[33] Hirata TT, Matsui HH. Freezing and Thawing Heat Transfer With Water Flow Around Isothermally Cooled Cylinders in Staggered and Aligned Arrangements. ASME. J. Heat Transfer.

1992;114(3):681–687. doi:10.1115/1.2911334.

[34] Yingxin, Z., Zhang, Y., 2001. Modeling of thermal processes for internal melt ice-on-coil tank including ice-water density difference, Energ. Buildings 33, 363–370.

[35] Vargas J. V. C. , Bejan, A., Dobrovicescu, A., 1994, The melting of an ice shell on heated horizontal cylinder, Transaction of ASME 116 702–708.

[36] Wu, S., Fang, G., Chen, Z., 2012. Discharging characteristics modeling of cool thermal energy storage system with coil pipes using n- tetradecane as a phase change material, Appl. Therm. Eng. 37, 336–343.

[37] Masahiko, Y., F. Shoichiro, K. Tsuyoshi, 2002 Performance analysis on the liquid- ice thermal storage system for optimum operation, Int. J. Refrig. 25 267–277.

[38] Tsuyoshi, K., Fukusako, S., Yamada, M., Itoh, K., 1999. Experimental on melting of slush ice in a horizontal cylindrical capsule, Int. J. Heat. Mass. Tran. 42, 2981–2990.

[39] Hirata T, Ishikawa M, Yamada K. "Crystal ice formation of solution and its removal phenomena on inclined cooled plate". *Int J Refrig*, 2002; 25:190–198.

[40] Sait, H.H. and A.M. Selim, Charging and discharging characteristics of cool thermal energy storage system with horizontal pipes using water as phase change material. Energy Conversion and Management, 2014. 77(Supplement C): p. 755–762.

[41] Sait, H.H. and H.B. Ma, *Experimental Study of Micro/Miniature Heat Pipe subject to High Heat Flux*, in *The fourth Saudi Technical Conference and Exehibit* 2006, TVTC: Riyadh, Saudi Arabia.

[42] Sanaye, S. and A. Shirazi, Thermo-economic optimization of an ice thermal energy storage system for air-conditioning applications. Energy and Buildings, 2013. 60: p. 100–109.

Chapter 6

News Aspects Theoric and Experimental to Paraffins Compounds

Eloi Alves da Silva Filho, Fabrício Uliana and Arlan da Silva Gonçalves

Abstract

The paraffinic compounds are important to new investigation on the properties physics and its correlation with theoric dates, because in literature no is completely clarified. However, there are some studies on the formalism for developing asymptotic behavior correlation for homologous series paraffin compounds. In this work is show that the effect of parameters theoric obtained by molecular modeling can be correlated with experimental dates. To paraffins as pure, for example, n-hexane, C_6H_{14}, MW 158 g/mol, is composed of two groups CH_3 and four groups CH_2 and its can depending of structure molecular ramification to predict what your dependency with thermodynamics data. Therefore, the molecular modeling of paraffinic compounds uses a methodology that looks for data correlated with the structure of the molecule complemented with experimental data. The objective this study is correlated this molecular data with some thermodynamics data as enthalpy of formation and other parameters.

Keywords: thermodynamics data, molecular modeling, paraffins compounds, theoric properties, molecular structure

1. Introduction

Paraffins compounds can be found in all phases of the petroleum production and very researched due to the industrial interest, mainly in what concerns the generation of energy and combustible material. Thermodynamic properties, such as heats of vaporization, specific heats, free energies, internal energies, entropies and enthalpies or interacting heat contents for petroleum hydrocarbons and their mixtures like paraffins compounds are very importance in chemical and work associated with petroleum refinery operations like other activities in this field. With the help of literature data and complementing with computational calculations of the thermodynamic properties of paraffins today it becomes possible to correlate theoretical data with experimental data [1].

In literature has been reported some study with a general equation for correlating the thermodynamic properties of n-paraffins and n-olefins, and other homologous series [2–6]. The recent work show that the modeling the thermal conductivity of n-alkanes via the use of density scalling for molecular liquids has proven to be a powerful estimation technique for transport properties [7]. It is important to

Molecular Formula	Name	Molecular Mass(u)	Boiling point b.p.(K)	Melting point m.p.(K)
CH_4	*Methane*	16	111.0	90.5
C_2H_6	*Ethane*	30	184.4	101.0
C_3H_8	*Propane*	44	230.9	85.3
C_4H_{10}	*Butane*	58	272.4	134.6
C_5H_{12}	*Pentane*	72	309.1	143.3
C_6H_{14}	*Hexane*	86	341.9	178.5
C_7H_{16}	*Heptane*	100	371.4	182.4
C_8H_{18}	*Octane*	114	398.7	216.2
C_9H_{20}	*Nonane*	128	423.8	222.0
$C_{10}H_{22}$	*Decane*	142	447.1	243.3

Table 1.
Physical properties of melting point and boiling point to Paraffins.

highlight that a database query can find many thermodynamic properties and are available for free on the internet, for example, NIST Chemistry WebBook [8], but no have data to interaction energy for paraffins. In preview work, Hadjieva and collaborators [9] showed that the increasing industrial interest in finding heat storage materials for efficient use of thermal energy in corresponding applications stimulates the investigation of phase change materials. Paraffin mixtures have been evaluated as suitable thermal storage materials (TSM) with melting points in the temperature range of 25–100 °C.

In general the paraffins compounds have little reactivity but it has a good interaction molecular due to being apolar structure and its directs some studies of its interaction energy ($\Delta_{int}E/kJ{\cdot}mol^{-1}$) when interactions occur between two n-alkenes such as pentane binding to another pentane where weak van der Waals interactions predominate. Therefore, the paraffins, the longer the chain (-CH2), the more interatomic sites there are for interactions between the molecules with a larger and stronger attractions or lower vapor pressures with higher boiling points. The literature [10, 11] shows that indeed boiling points for n-alkanes increase with increasing chain length. While often this trend is thought of as increasing boiling point with molecular wight (**Table 1**), it is really the increasing of the intermolecular forces that cause this not the increasing mass. The first four molecules, C1 to C4 are gases, C5 to C17 are liquids and those containing 18 carbon atoms or more are solids at 298 K.

The conformations of C2, ethane, is special, and the torsional strain and eclipsed form maximum torsional strain, it can be that rotation around C-C bond in is not completely free, and can affect the melting point.

In this work, the interaction energy, $\Delta_{int}E$ and formation enthalpy, Δ_fH were studied for five paraffins compounds like pentene, hexane, octane, undecane and tetradecane respectively, by computational calculate and correlation with experimental data, for evaluation of linearity found between experimental and theoretical data.

2. Methodology

2.1 Computational

The computational calculate in this work was used a machine with core processor i5-5200u 2.20GH and with 8G of RAM, 250D SSD, gforce 920 m video card running Xubuntu Core operating system (Linux distribution based on Ubuntu 16.04) and molecular modeling software distributed free of charge.

The work began with the construction of the geometry of the reagents used in the AVOGADRO program. In this program the graphical interface is used for the disposition of atoms and bonds in their proper places generating a three-dimensional representation of the molecules. The generated structures can be recorded in coordinate files that specify the positions of each atom in a Cartesian space (Cartesian coordinates) or indicate the bond length, angles and dihedral for each atom of the system (z-matrix) like is showed in **Table 2**.

The generated structure is then optimized so that the values of bond lengths, angles and dihedra are optimized in order to obtain geometries closer to those experimentally determined. To perform this optimization, one needs to use a set of parameter data for several atoms in several different chemical environments called the force field.

Once created and optimized, the system is saved to the PDB file. Using the program OPEM BABEL, the PDB file containing the reagents for the input files of MOPAC2016 was generated, thus generating a file <FileName> .mop.

In the .mop file the first line of the is intended for the insertion of the keywords that will determine what type of calculation the program should carry out the second line is a space to put comments that may be relevant and the other lines are the set of coordinates that representation the system. The following **Figures 1–4** show three-dimensional structure obtained from some paraffinic compounds used this computational methodology.

File	Edit	Format	Display	Help	
20					
	Energy:			−5.4744410	
C		−3.07778		0.34773	0.00881
C		−1.76950		1.11704	0.09877
H		−3.92678		1.02945	0.12005
H		−3.14344		−0.40822	0.79784
H		−3.17101		−0.15594	0.95864
C		−0.56337		0.19053	−0.05321
H		−1.72241		1.63457	1.06399
H		−1.74982		1.88544	0.68281
C		0.75059		0.96820	0.03782
H		−0.61754		−0.32775	−1.01837
H		−0.59011		−0.57856	0.72836
C		1.95672		0.04170	−0.11414
H		0.80476		1.48650	1.00297
H		0.77733		1.73729	−0.74377
C		3.26500		0.81102	−0.02427
H		1.90961		−0.47590	−1.07933
H		1.93706		−0.72666	0.66749
H		3.33064		1.56690	−0.81337
H		4.11399		0.12929	−0.13547
H		3.35824		1.31477	0.94314

Table 2.
Recorded in coordinate files.

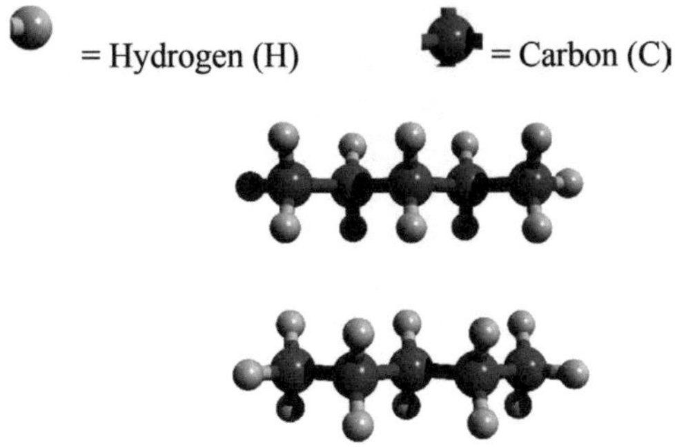

Figure 1.
Three-dimensional representation of the pentane molecule.

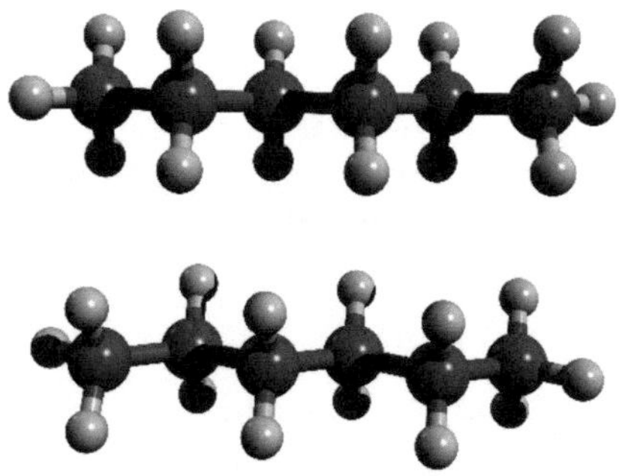

Figure 2.
Three-dimensional representation of the hexane molecule.

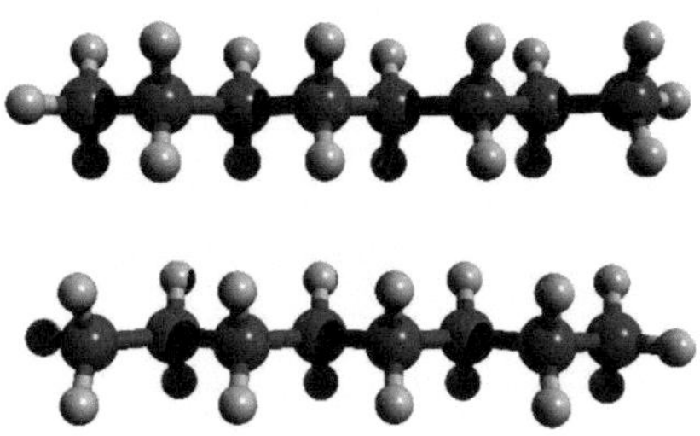

Figure 3.
Three-dimensional representation of the octane molecule.

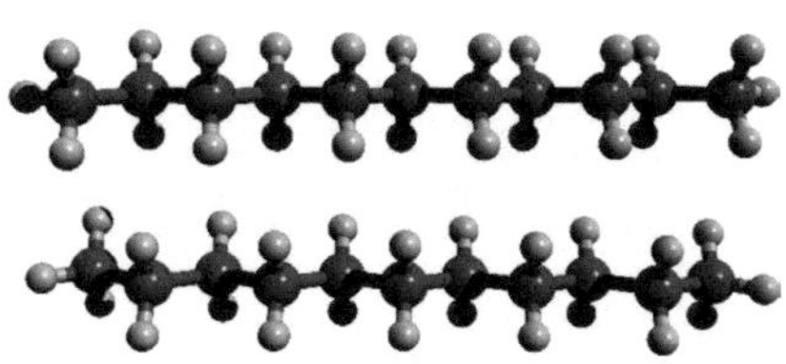

Figure 4.
Three-dimensional representation of the undecane molecule.

3. Results and discussion

The result of interacting energy, separate are in the **Table 3** and graphic in **Figure 5**, shows that a energy of interaction increasing with molecular mass, with a good correlation between each data, $r^2 = 0.99627$. This show that the interacting

Molecule	Interacting Energy ($kJ \cdot mol^{-1}$)	Separate energy ($kJ \cdot mol^{-1}$)	ΔintE ($kJ \cdot mol^{-1}$)	Boiling point Experimental (°C) [8]
Pentane	−295.48	−278.27	17.21	36.0
Hexane	−342.00	−320.16	21.84	68.71
Octane	−436.05	−404.45	31.60	125.59
Decane	−436.05	−532.31	44.69	196.21
Tetradecane	−718.54	−661.68	56.86	255.09

Table 3.
Values for energy of interaction between the molecules and the experimental boiling point.

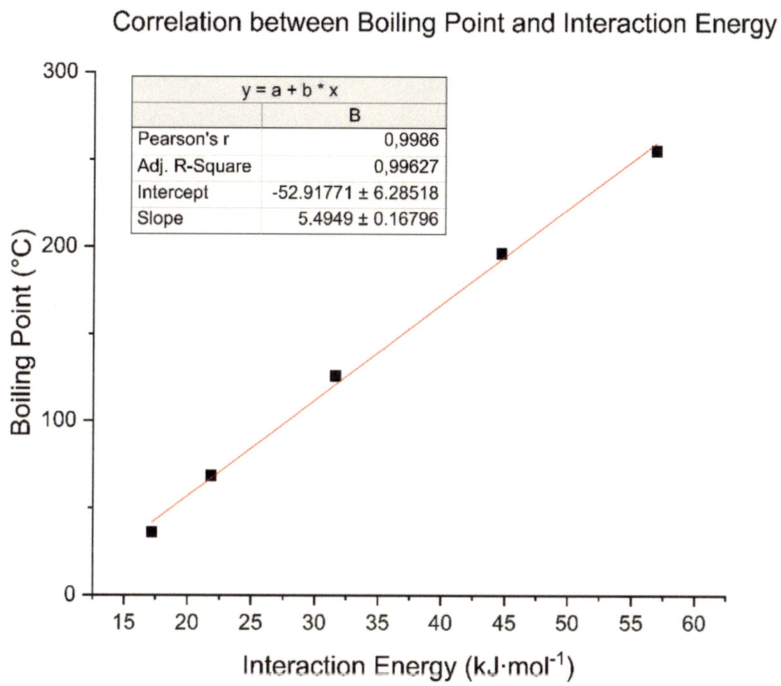

Figure 5.
Correlation between the calculated interaction energy and the experimental boiling point.

energy is a important parameter thermodynamic to explain the fact that intermolecular van der Waals increase with increase of the molecular size or the surface area of the paraffins molecule.

On the other hand the paraffins compounds can have infinite number of conformations by rotation around –C-C- single bonds and consequently generates of 1–20 kJ/mol in small energy barrier due weak interaction between adjacent bonds as called torsional strain. Then to obtain a stable molecule of paraffinic compounds it is necessary a more complete analysis of other parameters thermodynamics such as the enthalpy of formation ($\Delta_f H°$) that in this work was calculate using the program MOPAC201 where $\Delta_f H°$ were compared with experimental data, as can be observed in the **Figure 6**.

This formation enthalpy are very agreement until C_5H_{12} but the C_6H_{14} there is an increase in the difference that can be attribute a phase change and also in your properties physics as boiling point.

Thus, the ionization energy (IE) is also very important for understanding this linearity in **Figure 6** and have been investigated with results in **Figure 7**.

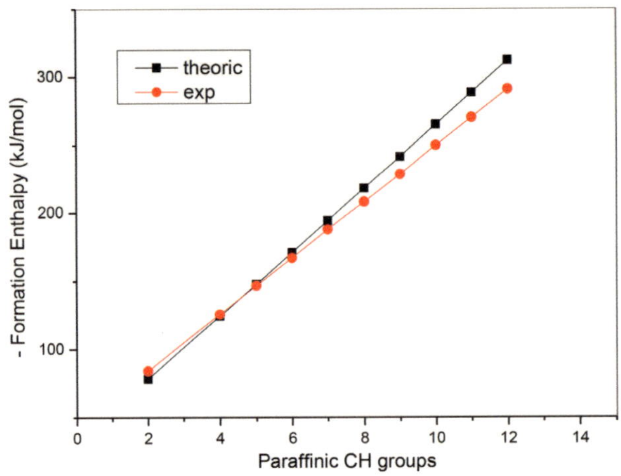

Figure 6.
Formation enthalpy experimental compared with theoric.

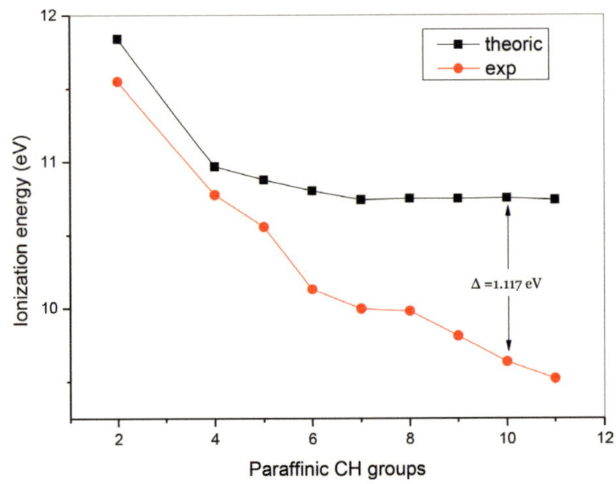

Figure 7.
Ionization energy experimental compared with theoric.

The ionization energy, also called ionization potential (IE) was defined according to IUPAC in 1994 like the minimum energy required to remove an electron from an molecular entity (its vibrational ground state) in the gaseous phase [12]. If the resulting molecular entity is considered to be in its vibrational ground state, one refers to the energy as the "adiabatic ionization energy". If molecular entity produced possesses the vibrational energy determined by the Franck Condon principle (according to which the electron ejection takes place without an accompanying change in molecular geometry), the energy is called the "vertical ionization energy". The ionization energy of a stable species, that is, any molecule that may exist, is always positive.

In the **Figure** 7, was observed that experimental data were higher than the theoretically obtained data, for example, Decane have a difference of the Δ – 1.117 and this discrepancies between computed and experimental results can be accounted for theoretical contributions, that are often based on the atomic groups or chemical bonds. This ionization energy also lowered by delocalization of charge as the molecular size is increased and related to some molecular property, common to paraffinic compounds.

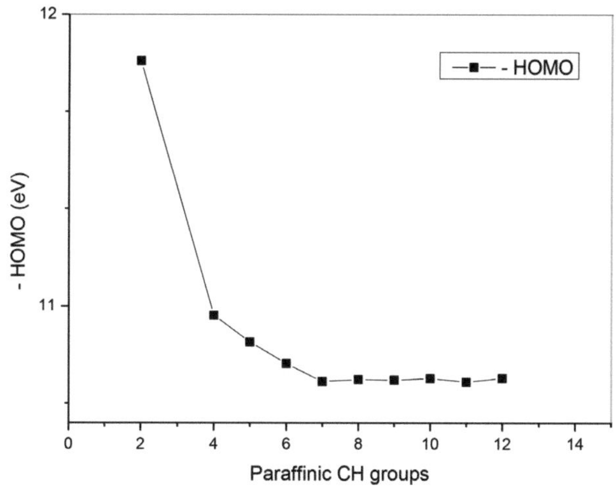

Figure 8.
HOMO energy orbital to CH paraffinic groups.

Hall [13] in earlier paper reported that the ionization potentials of some paraffinic molecules could be obtained of antisymmetrical molecular orbital localized over the CH bonds. In the **Figure 8**, was observed that to Paraffinic molecules of C7H16 to C12H26 are more stable to the ejection of an electron from the highest occupied molecular orbital (HOMO).

According of Cao and Yuan [14] reported that is more easily polarized the paraffinic molecule, the more stable the final-state (charged molecule), and furthermore, the lower the ionization potential will be. That is to say, the substituent polarizability effect can facilitate the ionization.

4. Conclusions

We conclude that the interaction energy, $\Delta_{int}E$ and formation enthalpy, Δ_fH were by computational calculate and correlation with experimental data, for to C_2 - C_{12} were important in this study due a good linearity found between experimental and theoretical data. With this model of computational calculate can be applied to other types of branched paraffinic compounds. The results of this study show that molecular orbital HOMO and polarizability are an mportant parameter to ionization energy of Paraffinic compounds.

Acknowledgements

This work was supported by the Federal University of Espírito Santo, Department of Chemisty and Coordination of Improvement of Higher Level Personnel (CAPES).

Conflict of interest

No.

Author details

Eloi Alves da Silva Filho[1*], Fabrício Uliana[1] and Arlan da Silva Gonçalves[2]

1 Department of Chemistry, Federal University of Espírito Santo, Vitória, ES, Brazil

2 Federal Institute of Education, Science and Technology of Espírito Santo, Vila Velha, ES, Brazil

*Address all correspondence to: eloi.silva@ufes.br

References

[1] Marano JJ, Holder GD. General Equation for Correlating the Thermophysical Properties of n-Paraffins, n-Olefins, and Other Homologous Series. Formalism for Developing Asymptotic Behavior Correlations. Ind. Eng. Chem. Res. 1997;36:1887-1894.

[2] Coutinho JAP, Ruffier-Méray V. Experimental and Thermodynamic modeling of paraffinic wax formation in undercooled solutions. Ind. Eng. Chem. Res. 1997;36:4977-4983.

[3] Marano JJ, Holder GD. A general equation for correlating the thermophysical properties of n-paraffins, n-olefins, and other homologous series. 3. Asymptotic behavior correlations for thermal and transport properties. Ind. Eng. Chem. Res. 1997;36:2399-2408.

[4] Jin Y, Wunderlich B. Heat capacities of paraffins and polyethylene. J. Phys. Chem. 1991:95:9000-9007.

[5] Pellegrini LA, Gamba S, Bonomi S., Calemma V. Equilibrium constants for isomerization of n-paraffins. Ind. Eng. Chem. Res. 2007;46:5446-5452. DOI: 10.1021/ie0705204

[6] Hunter KC, East ALL. Properties of C-C bonds in n-alkanes: relevance to cracking mechanisms. J. Phys. Chem. A 2002:106:1346-1356.

[7] Macías-Salinas R. Modeling the thermal conductivity of n-alkanes via the use of density. J. Chem. Thermodynamics 2018:116:363-371.

[8] National Institute of Standards and Technology. 2018. Available from: https://webbook.nist.gov/chemistry/ [Accessed:2018-12-09]

[9] Hadjieva M, Kanev S, Argirov J. Thermophysical properties of some paraffins applicable to thermal energy storage. Solar Energy Materials and Solar Cells 1992:27:181-187.

[10] Yuqing Q, Valeria M. Strength of Alkane–Fluid Attraction Determines the Interfacial Orientation of Liquid Alkanes and Their Crystallization through Heterogeneous or Homogeneous Mechanisms. Crystals 2017: 7: 2-18. DOI:10.3390/cryst7030086

[11] Kostis M, Dimitrios T. Thermophysical properties of n-alkanes from C1 to C20 and their prediction for higher ones. Fluid Phase Equilibria, 1990:56: 119-140. DOI: 10.1016/0378-3812(90)85098-U

[12] Muller P. Glossary of terms used in Physical Organic Chemistry. Pure & Appl. Chem., 1994:66:1077-1184. DOI: 10.1351/pac199466051077

[13] Hall GG. The ionization potentials of some paraffinic molecules. Transactions of the Faraday Society, 1954:50:319-322. DOI: 10.1039/tf9545000319

[14] Cao C, Yuan H. On Molecular Polarizability. 4.Evaluation of the Ionization Potential for Alkanes and Alkenes with Polarizability. J. Chem. Inf. Comput. Sci., 2002:42:667-672. DOI: 10.1021/ci010344o